D1131613

SPACE PROBES

A Firefly Book

Published by Firefly Books Ltd. 2011

Copyright © Firefly Books Ltd.

Originally published as *Histoire visuelle des sondes spatiales*
Copyright © 2010 Éditions Fides, Quebec

All rights reserved. No part of this publication may be reproduced, stored in a retrieval system, or transmitted in any form or by any means, electronic, mechanical, photocopying, recording or otherwise, without the prior written permission of the Publisher.

First printing

Publisher Cataloging-in-Publication Data (U.S.)
Séguéla, Philippe, 1959-
   Space probes : 50 years of exploration from Luna 1 to New Horizons / Philippe Séguéla.
Originally published as Histoire visuelle des sondes spatiales; Montrel: Éditions Fides, 2009.
[376] p. : ill. (some col.), col. photos., charts ; cm.
Includes bibliographical references and index.
Summary: The history of space probes, from Luna 1 in1959 to New Horizons in 2006, and updates on probes planned for 2011 to 2025. Each probe is described in terms of objectives, technology, challenges, information gained and lessons learned. Photographs and technical drawings give a view of each mission, and special features focus on engineers, physicists and astronomers.
ISBN-13: 978-1-55407-944-5
1. Space probes. 2. Outer space -- Exploration. I. Title.
629.435 dc22   TL795.3.S458 2011

Library and Archives Canada Cataloguing in Publication
Séguéla, Philippe, 1959-
   Space probes : 50 years of exploration from Luna 1 to
New Horizons / Philippe Séguéla.
Issued also in French under title: Histoire visuelledes sondes spatiales.
Includes bibliographical references and index.
ISBN 978-1-55407-944-5
1. Space probes. 2. Outer space--Exploration. I. Title.
TL795.3.S4313 2011   629.43'5   C2011-902382-2

Published in the United States by
Firefly Books (U.S.) Inc.
P.O. Box 1338, Ellicott Station
Buffalo, New York 14205

Published in Canada by
Firefly Books Ltd.
66 Leek Crescent
Richmond Hill, Ontario L4B 1H1

Art editor: Gianni Caccia
Production: Carole Ouimet
Graphics and image processing: Bruno Lamoureux
English translation by Klaus and Margaret Brasch

Printed in China

Philippe Séguéla

SPACE
PROBES

50 YEARS OF EXPLORATION FROM LUNA 1 TO NEW HORIZONS

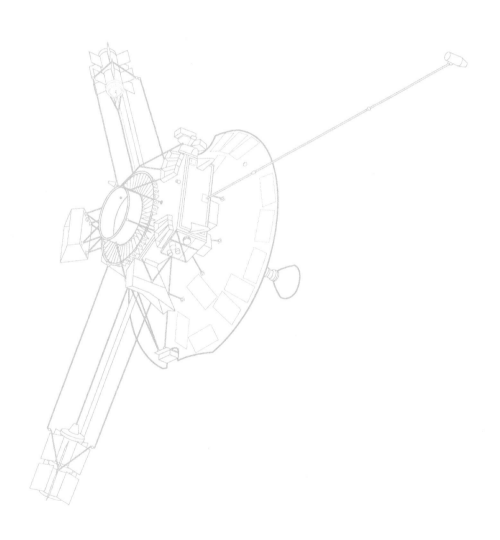

FIREFLY BOOKS

TABLE OF CONTENTS

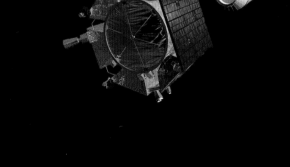

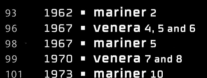

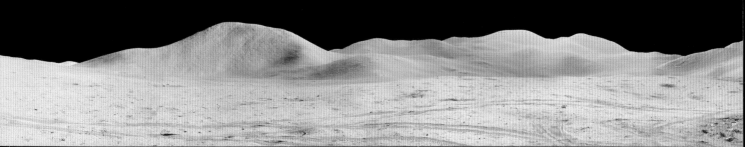

FOREWORD

AS HUMAN SPACECRAFT now close in on the last unreached physical frontiers of the solar system (namely Pluto and the surface of a comet), I am reminded of Carl Sagan's poetic exaltation of the unique place our generation holds in the entire parade of human exploration. Sagan explained that until our time, people could speculate idly about conditions on other worlds but never hope to discover them, but after us, anybody with the curiosity to ask will be able to get the answers from historical records stored on some handy nanochip. We alone of all humans were present at the transition, at the brief span of time when we converted the worlds of the Solar System from "mostly unknown" to "mostly known."

This fine book is a tribute to that exploration explosion, as our robot emissaries reached, one by one, farther and farther goals. Its visual value is superb, and the text is a perfect accompaniment to the prettiest pictures I've ever seen collected under one set of covers. It offers a narrative of what we have discovered with space probes, with attention paid to how we accomplished it. And it does so with a commendable editorial discipline of not wandering too far astray, except where occasional side notes add to the sense of discovery. This approach perfectly communicates how, even though our far-flung senses and manipulators may be mechanical, their guiding spirit is eminently human, and the exploration instinct these automata serve is based on the human agenda, not on theirs.

It is also breathtakingly visual, especially in the images from more recent explorations. The production qualities made me run my fingers over many of the pages to make sure they weren't really as three-dimensional as they looked.

This book is not a nostalgic retrospective of a completed, finished task. It is instead a survey of the first of what will be many ascending waves of reconnaissance. As such, it is a guidebook to off-Earth theatres of astonishment that await our year-by-year more capable automata and, eventually, ourselves.

We are realizing that Sagan's flowery words may have underappreciated the time scale of this sudden explosion of knowledge that we have been experiencing throughout our lifetimes. Young readers need not be envious of the uniqueness of their elders' extraterrestrial epiphanies, because I suspect the best is yet to come. The knowledge explosion is far from over.

Sagan himself could scarcely have dreamed — or hoped to discover — what we are on the verge of finding out: that our Earth-bound, narrow-minded fixation on "life as we know it" doesn't even apply to our home planet. In only a fraction of a lifetime we have come to realize that most living terrestrial organisms inhabit ecological niches away from sunlight, under thermal and chemical conditions that we now realize also occur inside the skins of at least two dozen other worlds. And we have realized that spores can survive transplanetary trajectories, impelled by natural asteroid impacts, which destroys any notion of biological quarantine all the way back to the birth of the Solar System.

If we reason by analogy with what we now know about the largest component of life on Earth — the subsurface microbes — then we have a good basis to suspect that "life" in similar environments on other worlds would also be microbial, not multicellular. But that kind of logic, assuming that "life as we know it" is the limit of nature's creativity, is the same discredited formula that led us to be so surprised by the current new paradigm of potential — even probable — enclaves of water-based "life" elsewhere in the solar system. When fundamental physical and environmental factors are significantly different — pressure, gravity, radiation, whatever — what we call "life" can be expected to show remarkable adaptational flexibility. But whatever we "expect," we should certainly expect to be astonished again and again, and maybe from directions that we least "expect" it! The only remedy is to go and look, go and dig, go and poke around — and we will.

We are finally ready for the curtain to rise on "Space Reconnaissance 2.0," as we begin to catalog the biochemical provinces of what we once had convinced ourselves was a dead and sterile outer-space outback. Now at the early dawning of that endeavor, this book is a grand tribute to what has been found out so far and to the people who did it, and moreover, it is a testament to the impulses that will power the next stage of exploration and discovery.

The last sentence of the author's narrative is this: "The discovery of life elsewhere in the solar system or beyond will be a defining moment in human history and probably shift future space exploration in directions impossible to appreciate today." But even though it's obvious, it bears saying again and again. Making it the book's last sentence allows it to serve as the jumping-off point for the next retrospective, to be written half a century from now, based on the astonishments awaiting the next generation — who will NOT be disappointed.

We must be reminded that even Sagan, who was mind-bogglingly poetic, was very likely to have been wrong to think his generation would ever monopolize history's most spectacular space discoveries. The past, as another poet said, is not the epilogue to a vanished age of heroes. The past, particularly in planetary exploration as chronicled here, is only prologue.

JAMES E. OBERG

AUTHOR'S NOTE

Through a happy coincidence, I was born in 1959, the year the very first space probe, Luna 1, was launched. Now I am a professor at McGill University, where I direct a team of molecular neuroscientists at the Montreal Neurological Institute. What's the link between neuroscience and astronautics, you ask? Foremost is my interest in investigating the infinitely small and the infinitely large: from the microscopic to the cosmic realm. Whether I'm studying cells or planets, I am curious about new frontiers and discoveries that provide us with a clearer vision of our place in the Universe. But there is another important connection: the evolution of the "brains" and "sensory organs" of space probes. The simple relay switches on the first Soviet probes have given way to veritable nervous systems on current spacecraft, with powerful computers linked to high-resolution digital retinas as well as other advanced receptors.

My first encounter of the "third kind" with astronautics happened in 1977, while I was a biology student in Bordeaux. In a course on audiovisual methods I volunteered to produce an animated short movie about the Viking mission and its first tentative efforts at looking for extraterrestrial life (exobiology) on Mars. By way of introduction, the physics professor in charge of the course provided us with original photographs of Mars, transmitted by the Mariner 9 probe. He had obtained them from a colleague working at the Jet Propulsion Laboratory in California, the Mecca of robotic missions. Holding in my hands original photographs of another world, 150 million kilometers from Earth, was a seminal experience for me. Since that time, I have kept myself informed on advances in space-probe technology. Not finding any comparable books in either public or university libraries, I decided to write a history of space exploration in order to share my passion.

This book is an invitation to take a voyage, a voyage that began toward the end of the 1950s with the first excursions into outer space. Technology has advanced enormously since the start of the space age, and a revolution has taken place in communications — from the rapid growth of television to the Internet. Information released on NASA and ESA websites is quickly picked up by press agencies and TV networks. Who has not heard about Voyager or Pathfinder? Thanks to worldwide communications systems, we can now receive, at home, "live" images sent by spacecraft currently operating in different parts of the solar system.

Through the means of this book, I too provide a voyage through space, from the Moon to Pluto. There is still so much to discover in order for us to understand the birth and evolution of the solar system, and space probes continue to flood us with information. As a member of the Voyager mission pointed out, we learned more about Saturn in a week than during all prior human history. For scientists and the public alike, the images of extraterrestrial landscapes are the most spectacular aspects of these missions. In addition to their graphic beauty, such images fire our collective imagination and feed our dreams.

Because of the high quality of these images, readers can see for themselves how far we have advanced technologically since the early efforts of 1959. This book pays homage to all those who have helped make these missions a reality. Let us not forget that despite all the risks, these probes are sent into space at dizzying speeds and exposed to extreme temperatures and intense radiation while they hurtle toward inhospitable, moving targets at "astronomical' distances from us. Mistakes are not forgiven in outer space, and despite years of careful planning, many missions have failed. With this in mind, I have explained the technical aspects of the various missions as well as their trajectories, so as to showcase the ingenuity and expertise of their designers.

By balancing the visual and technical aspects of these missions, I have written this book for the enjoyment of the young and not so young, and for all those readers who want to retrace, at their leisure, the first 50 years of our exploration of the solar system.

In recognizing the long road traveled during this first half-century of solar-system exploration — a truly pioneering effort — this book also hopes to inspire others to pursue careers in astronautics and astronomy. If that happens, this book and its author will have realized their goal.

PHILIPPE SÉGUÉLA

Laying the groundwork for solar-system exploration

Definition of a space probe: An uncrewed spacecraft, freed of the Earth's gravitational force, whose purpose is to explore one or more solar-system objects.

NOWADAYS, THE ARRIVAL OF A SPACE PROBE on Mars seems to be almost a routine event in the eyes of the public, who expects to follow its progress on television and the Internet. It's hard to believe that some 50 years have elapsed since the launch of the first space probe, Luna 1. This time frame is misleading, however, as planetary exploration has had a long, costly and exciting history. It requires enormous expertise, patiently obtained under constant questioning and scrutiny, which is based as much on science and technology as operational coordination. The 20th century will be remembered as the beginning of space exploration, yet in 2011, only a small number of nations is capable of launching interplanetary spacecraft, an astronautical accomplishment far more difficult than placing a satellite into orbit. Due to the high cost of solar-system exploration, the history of space missions reflects a continuing struggle between scientists and politicians to secure the necessary financial commitment to realize each mission. Leaders and taxpayers of even the wealthiest nations have to strike a balance between pure scientific curiosity and more urgent and concrete matters here on Earth.

Why should we undertake exploration of the solar system?

From a cosmic perspective, the solar system is our birthplace. Thanks to robot space probes, we now have the means to venture increasingly farther from our home planet. Our curiosity is big and our scientific objectives are many. Space exploration is also driven by profound questions about our place in the Universe, and bridges scientific inquiries pertaining to human civilization, from biology to astronomy. How did the solar system originate

OPPOSITE PAGE

The American inventor, Robert H. Goddard, with the world's first liquid-fueled rocket, launched in March 1926.

and evolve? What were the conditions that fostered the emergence of life on Earth? Has life appeared elsewhere in the solar system? What resources and dangers could impact further human expansion into space?

The first 50 years of planetary exploration have been full of surprises. The solar system has presented us with an enormous variety of lunar and planetary landscapes. No one could have imagined, for example, that robotic spacecraft would find a gigantic canyon on Mars, methane lakes on Titan, sulphur volcanoes on Io or an ice-covered ocean on Europa. Despite this great geologic diversity, not a single trace of life has been detected so far. When the New Horizons mission arrives at Pluto in 2015, our entire solar-system village will have been visited. After going to our neighbors, Venus and Mars, we have learned that conditions on Earth are totally unique in the solar system, conducive to the appearance and continuous evolution of life for some four billion years. In the coming half century, the exploration of those worlds where some form of life, past or present, may exist, will no doubt be as exciting an astronautical adventure as the preceding one. The discovery of any type of extraterrestrial life forms would be an extraordinarily important finding, with profound and unprecedented philosophical implications.

In addition to satisfying our intellectual curiosity, space exploration provides many benefits difficult to quantify, but real nonetheless. Mistakes can be fatal in space, where there is no second chance. The manufacture of a robotic spacecraft and its launch vehicle have to be carried out with maximum precision and the best materials, and then rigorously tested to insure it will operate under extreme conditions. Such high standards were largely unknown before the space age. All the laboratories and subcontractors involved in a given space mission must meet these criteria. Industry and universities have met the challenge to produce higher quality research and materials, and society as a whole has benefited. In the longer term, space exploration has provided insight into raw materials and energy sources that will extend our economic sphere of influence into the solar system. Such direct material benefits of space exploration will be of immense consequence to future generations. This fact has not escaped emerging nations like China and India, who are actively engaged in developing their own national expertise in interplanetary exploration.

Apart from the practical reasons for the exploration of space, the subjective aspect for doing so is no less important. National pride and the drive to be the first and the best, as was the case during the Cold War, will continue to be major considerations. Equally noteworthy is the profound human desire to leave a lasting legacy for future advanced civilizations. Just like the pyramids and great cathedrals, solar-system space probes represent glorious monuments to human ingenuity. There is no doubt that those first steps into space will be seen as a source of inspiration, and might well be the sole surviving evidence of our existence, should our species annihilate itself.

How are space probes launched?

In order to attain their goal and accomplish their mission, space probes must escape the Earth's gravity. This has been a human dream since the earliest times. Various fictional means have been invoked to attain that dream, using balloons, flying machines, harnessing birds or other magical methods. However, Newton's laws have constrained such efforts in a simple equation:

$$V_e = \sqrt{\frac{2GM}{r}}$$

Where V_e is the escape velocity, G is the universal gravitational constant, M the mass of the planet, and r the radius of the planet.

For the Earth, escape velocity is 11.2 kilometers per second: 34 times the speed of sound (Mach 34), or 10 times faster than a bullet fired from a rifle!

The Earth rotates about its axis at a speed of 465 meters per second at the equator. Therefore a rocket launched eastward tangentially to the equator only requires a velocity of 10,735 (= 11.2 – 0.465) kilometers per second to reach escape velocity, while one launched in a westerly direction must reach 11,665 (= 11.2 + 0.465)) kilometers per second. The Earth's rotational velocity varies as a function of the cosine of a given latitude, so most space launching stations are located as close as possible to the equator. The primary American launch facility, Cape Canaveral, is located at latitude 28°28′ N in Florida, while the European space center at Kourou in French Guyana lies at 5°14′ N.

It was not until the end of the 19th century, through authors like Jules Verne and H. G. Wells, that the notion of space travel was popularized in a rational fashion. Rational — but not realistic. In *From the Earth to the Moon*, Jules Verne proposed an enormous cannon firing a shell that was modified

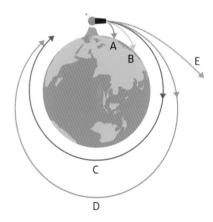

Isaac Newton (1643–1727) imagined a cannon atop a very high mountain, firing projectiles (A–E) with ever increasing force. Projectiles A and B fall back to Earth, while C attains a circular orbit and D an elliptical orbit. Only E is launched with enough force to attain escape velocity and free itself of Earth's gravity.

as a spacecraft. In reality, this method faces two insuperable obstacles. First, the initial thrust needed to fire the projectile would generate enough energy to destroy the entire machine and, of course, its human cargo. Second, the maximum velocity attained by such a cannon shell (2 kilometers per second) is not enough to escape Earth's gravity. To overcome this problem, H.G. Wells postulated a new metal, cavorite, in his novel *The First Men on the Moon*. This imaginary metal, which had anti-gravity properties, was used as the outer shell of his spaceship.

The idea that rockets were key to the dream of space travel took hold at the turn of the 20th century. The first clear descriptions of the use of rockets date back to China in the 11th century, where they were used to repel attacks by the Mongols. Rockets were developed by the Chinese shortly after their invention of gunpowder, the explosive material needed to propel them. This technology was brought to the western world at the time of Genghis Khan, during his conquest of parts of Russia and central Europe. British officer and inventor Sir William Congreve (1772–1828) perfected artillery rockets, which were used during the Napoleonic wars. Rockets were mainly psychological weapons, since they were not accurate enough to really affect the outcome of a battle.

The three founding fathers of space flight

Rockets were little more than minor weapons and devices used for fireworks when the solitary and self-taught Russian author, Konstantin Tsiolkovsky (1857–1935), published the first theoretical work on space travel. *The Exploration of Cosmic Space by Means of Reaction Devices* was published in 1903. In it he described the principles and plans for a liquid-fueled rocket powerful enough to attain escape velocity of 11.2 kilometers per second. Ahead of his time, Tsiolkovsky described a liquid-fuel mix consisting of oxygen and hydrogen. Trajectory guidance by mobile elements in the gas jet and gyroscopic stabilization are also included in this book. At the late age of 60, the ingenious Tsiolkovsky introduced the principles of multistage rockets for crewed space missions and the concept of orbiting stations, both remarkably visionary ideas.

Working independently on the other side of the Atlantic Ocean, American physicist Robert H. Goddard (1882–1945) also became convinced that a fueled rocket could reach the upper atmosphere and outer space. Unlike the theoretician Tsiolkovsky, Goddard was an experimenter and, on March 16, 1926, his first liquid-propelled rocket was launched.

Robert H. Goddard

Konstantin Tsiolkovsky

Hermann Oberth

It reached all of 13 meters in elevation, which was enough to prove the validity of the concept. Unfortunately, the value of Goddard's pioneering work was not appreciated until after his death.

Around this same time period, influenced in his youth by the science fiction novels of Jules Verne and H.G. Wells, German student Hermann Oberth (1894–1989) independently devised the fundamental equation of propulsion reaction that had also been devised by Tsiolkovsky:

$$\Delta v = v_e \ln \frac{m_0}{m_1}$$

Where Δv is the change in velocity, v_e the ejection velocity of the gas, ln the natural logarithm, m_0 the initial mass of the spacecraft before ignition, and m_1 the total mass of the spacecraft at the end of ignition.

Oberth enjoyed a great deal of popularity in 1923 following the publication of his thesis, *Rockets in Interplanetary Space*, which was considered highly revolutionary at the time. In 1928 he served as consultant to cinematographer Fritz Lang for the production of his movie *Woman in the Moon*. They planned to launch a rocket at the premiere of the movie in 1929, but it didn't work out. Nonetheless, the concept of a liquid-fuel rocket, which Oberth tested successfully in 1935 for the Rumanian army, was taken up by the Germans, who hired him to work on the V2 missile.

From concept to reality

The works of Tsiolkovsky, Goddard and Oberth demonstrated that, compared to other types of vehicles, rocket motors had three distinct characteristics particularly well suited for space travel. First, gas ejection provided increasing acceleration, while the weight of the spacecraft gradually diminished. Second, the principle of action and reaction displacement was equally valid in a vacuum. Third, since combustion is internal, it can occur in interplanetary space, where ambient oxygen is not available. Rocket clubs formed by young enthusiasts soon appeared in many countries, including the VfR *(Verein für Raumschiffahrt,* "Society for spaceflight") in Germany in 1927 and GIRD *(Grouppa Izoutcheniia Reaktivnovo Dvijeniia,* "Group for the study of reaction propulsion"), founded in 1931 in the Soviet Union. Hermann Oberth and Wernher von Braun (see biography on page 23) tested a number of rocket prototypes at the VfR, while Sergei Korolev (see

The Hohmann Trajectory

The principles of the celestial mechanics required for an interplanetary trip were described in 1925 by the German engineer Walter Hohmann (1880–1945) in his book, *Die Erreichbarkeit der Himmelskörper* (*The Attainability of the Celestial Bodies*). They are still applied today to calculate transfer orbits for spacecraft, in order to use as little fuel as possible.

The elliptical Hohmann trajectory (2) is tangential to the original orbit (1), leading to the desired orbit (3). Two rocket firings (ΔV and $\Delta V'$) are both necessary and sufficient to achieve that.

Using the Hohmann maneuver, it takes five hours to transfer a satellite from a low to a geosynchronous orbit, five days to transfer from there to lunar orbit, and 259 days to send a probe from Earth to Mars. In order to save time, a trajectory involving a gravitational slingshot is preferred to reach the outer planets.

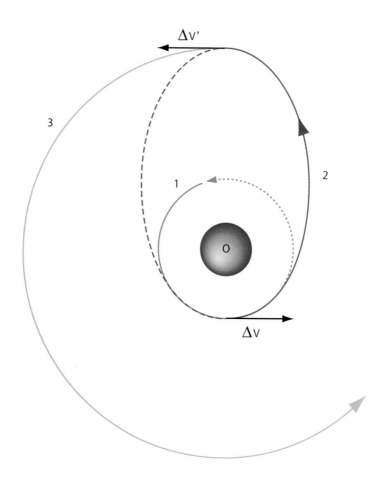

biography on page 21) led GIRD. The military quickly realized the potential these new vehicles provided for launching bombs and soon took control of operations. After attaining power in 1933, the Nazi Party forbade all civilian experimentation with rockets, and GIRD was quickly taken over by the Soviet Army in order to develop rocket engines for airplanes. The German army built an experimental rocket base at Peenemünde on the shores of the Baltic Sea, which led to production of "vengeance weapons" (*Vergeltungswaffe*), namely the V1 and V2. Many members of GIRD were

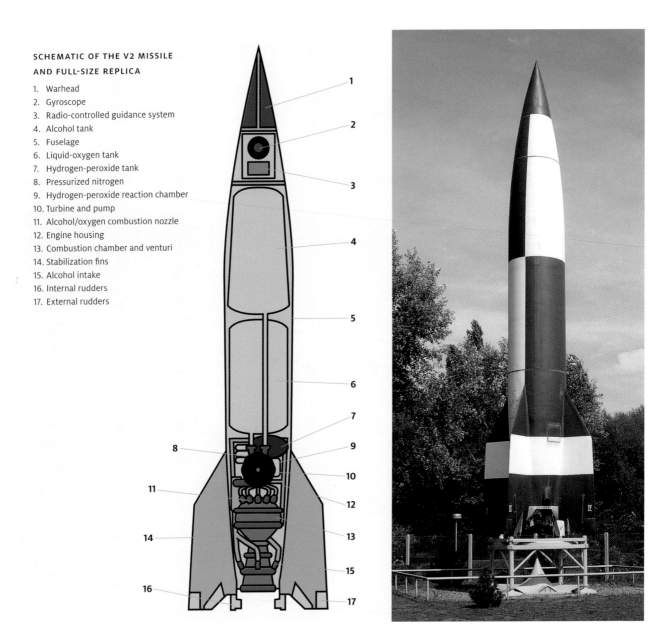

**SCHEMATIC OF THE V2 MISSILE
AND FULL-SIZE REPLICA**

1. Warhead
2. Gyroscope
3. Radio-controlled guidance system
4. Alcohol tank
5. Fuselage
6. Liquid-oxygen tank
7. Hydrogen-peroxide tank
8. Pressurized nitrogen
9. Hydrogen-peroxide reaction chamber
10. Turbine and pump
11. Alcohol/oxygen combustion nozzle
12. Engine housing
13. Combustion chamber and venturi
14. Stabilization fins
15. Alcohol intake
16. Internal rudders
17. External rudders

SERGEI PAVLOVITCH KOROLEV

(1907–1966)

Born in Jytomir in the Ukraine, Korolev was trained as an aeronautical engineer and became the chief architect of the Soviet space program during the 1950s and 1960s. He is credited with producing the R-7 "Semiorka" rocket, the launch of the Sputniks (see below) and the first space probes: Luna, Mars and Venera. In addition, he was instrumental in directing the first crewed Vostok flights (first human in space) and Voskhod (first extra-vehicular activity). A victim of Stalinist purges, he was imprisoned in 1936 and spent several months in the Siberian gulag before being transferred to a camp to work under the celebrated aeronautical engineer Andrei Tupolev. After finally being released, he was appointed a Colonel in the Red Army in 1945 for his remarkable work on jet-fighter aircraft. Subsequently,

he participated in the development of long-range nuclear missiles based on German V2 rocket plans. Due to the abuse suffered in prison camps, his health was fragile and he died prematurely at the age of 59 while undergoing surgery. His name was never officially announced for security reasons during the Cold War, and he lived unrecognized despite his many successes. It was only several decades after his death that Soviet authorities publicly acknowledged Korolev's major role in the history of astronautics.

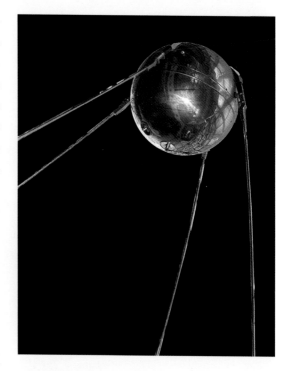

"The road to the stars has been opened."

eliminated during the Stalinist purges of 1938, but Korolev miraculously survived. After the war he produced some of the intercontinental ballistic missiles. The R-7 Semiorka rocket successfully launched in August of 1957. Capable of delivering a nuclear bomb weighing several tons at a distance of 7,000 kilometers, it changed the Cold War completely. With a vehicle like the R-7, Korolev's dream of space flight was finally realized.

Along with Valentin Glouchko and Mikhail Tikhonravov, he was subsequently given carte blanche to build the Soviet space program. The fear of being beaten by the Americans led them to launch Sputnik a few weeks later. On October 4, 1957, they took the whole world by surprise. One month after that, the launch of Sputnik 2, with the dog Laika aboard, marked the first time that a living creature was placed into orbit. With the help of Wernher von Braun and his team of renegades, the Americans consoled themselves with the launch of their first satellite, Explorer 1, placed into orbit in February of 1958. Only one year later, the Soviets went from orbiting a satellite to launching the first lunar probe, Luna 1 (see page 27), in 1959. The Americans had to wait a few months more for their turn to venture into interplanetary space, using a Juno rocket to launch the modest probe Pioneer 4 (see page 30).

From left to right, William H. Pickering, director of the Jet Propulsion Laboratory, James A. Van Allen, designer, and Wernher von Braun celebrate the first successful launch of an American satellite, Explorer 1, on February 1, 1958.

WERNHER VON BRAUN

(1912–1977)

This German engineer and physicist is one of the 20th century's prominent figures for his crucial role in the development of rocket technology, both in Germany and the United States. From 1920 to 1930, von Braun directed development of military rockets for the German army. During the Second World War, he joined the Nazi Party. From 1939 to 1942, he

directed operations at Peenemünde, the experimental rocket base dedicated to the development of the first cruise missile, the V1, and the first ballistic missile, the V2. In 1943, Hitler ordered ramped-up production of the V2, resulting in forced labor by concentration-camp prisoners from Dora-Mittelbau and Buchenwald. Working under such inhumane conditions caused more death among prisoners than by the missiles themselves. Von Braun's participation in this undertaking caused much controversy and tarnished his reputation. With the Soviet army approaching in 1945, von Braun and a large number of his collaborators surrendered to the Americans. Under a secret operation called Paperclip, he was transferred to the United States, where he helped develop guided missiles based on the V2 model. In 1960, he was appointed director of NASA's Marshall Space Flight Center, where he conceptualized and designed the Saturn rocket. The construction of the titanic Saturn 5, which launched the Apollo missions, was one of his greatest accomplishments. Because of this, he is considered the father of the American space program and a key player in its success in winning the race to the Moon.

Regarding the devastation caused by the first V2s in London, von Braun declared: "The rocket worked perfectly, except for landing on the wrong planet."

First Objective: The **Moon**

THE MOON is the sole natural satellite of Earth. It is also the only astronomical body visited by humans. Its changing phases and variable luminosity have always provoked curiosity and inspired artists. The Moon has the odd trait of being able to turn on its own axis in exactly the same time (27.32 days) that it executes a single rotation around the Earth. By doing so, it always displays the same face to Earth-bound observers. The hidden face of the Moon was photographed for the first time by the Soviet space probe, Luna 3, in 1959. The lunar surface is covered by a dusty layer called regolith, which varies in thickness from 3 meters in the basins to 20 meters in the high plateaus.

The Earth–Moon system: a double planet

The Moon is one of the largest satellites in the solar system and the Earth–Moon system is considered to be a double planet. The Moon's origin is still the subject of active debate in scientific circles. The current most accepted hypothesis is that there was a giant impact between the young Earth and a smaller planet, Theia, which would have occurred some 4.45 billion years ago, slightly after the birth of the solar system. This collision is thought to have ejected a colossal mass of matter into space, which then formed the Moon via a process of accretion.

Diameter: **3,475 km**
Mass: **0.012 [Earth = 1]**
Average Distance from Earth: **384,400 km**
Rotational Period: **27.32 days**
Orbital Period: **27.32 days**
Day Length: **29.53 days**
Average Diurnal temperature: **107°C**
Average Nocturnal Temperature: **-153°C**
Surface Gravity: **0.166 g**
Escape Velocity: **2.37 km/sec**

Nautilus and the Moon

The gravitational pull of the Moon explains the phenomena of tides on Earth. The ocean waters are affected by lunar gravity as the Earth rotates, and the daily ebb and flow causes friction of water masses against the ocean floor. The resultant loss of energy produces two results: the Earth's rotation gradually slows down, and the Moon distances itself very slowly from the Earth, at about 4 centimeters per year. The nautilus, a cephalopod mollusk with a spiral shell, has inhabited global seas for hundreds of millions of years. At each full moon, it rises to the water's surface and forms a new interior chamber. It has been observed that the older the fossilized shells, the more interior chambers can be found, thereby indirectly confirming the progressive distancing of the Moon and the elongating of the lunar month.

The Moon and space exploration

Due to its proximity to Earth, the Moon has been and remains the ideal target of all nations willing to test their technical expertise in the domain of space exploration. As far back as 1958, the Americans started the ball rolling by attempting the first lunar flyby. Unfortunately, all three attempts, Pioneer 1 (launched on October 11), Pioneer 2 (November 8) and Pioneer 3 (December 6), resulted in failures. It was left to the Soviets to truly inaugurate lunar exploration, by means of space probes, with their launch of Luna 1 on January 2, 1959. The race to the Moon between the Americans and the Soviets during the 1960s resulted in the launch of several observational probes and finally in Neil Armstrong's first step on the Moon on July 20, 1969, during the Apollo 11 mission. All in all, 12 men have walked on the Moon. In 2008, NASA proposed a new lunar exploration program, with the goal of constructing a permanent base on the Moon. For budgetary reasons, the program was cancelled, but the base would have facilitated the launching of crewed missions to Mars and beyond. Large quantities of water in the form of ice, from past cometary impacts, have been found in certain polar craters that have never been exposed to sunlight (see page 89). Utilization of such resources will require mechanisms of extraction and purification. The question of water on the Moon is, of course, of great importance with respect to the establishment of future human colonies. ☾

Luna 1 and 2

O N JANUARY 2, 1959, at the Baikonur Cosmodrome in Kazakhstan, a more powerful version of the R-7 Soviet rocket, the Vostok 8K72, carried Luna 1 toward the Moon. The spherical Luna 1 probe measured 62 centimeters in diameter, weighed 361 kilograms and carried five scientific instruments. By reaching Earth escape velocity, Luna 1 is now recognized as the first artificial object to free itself from the planet's gravitational pull and the first interplanetary space probe in history. On January 3, 1959, Luna 1 released sodium gas 113,000 kilometers from Earth, in order to create an artificial comet. Visible above the Indian Ocean through its brilliant orange tail, the sodium-cloud path was followed by many astronomical observatories.

On January 4, 1959, Luna 1 passed within 5,995 kilometers of the Moon's surface after a 34-hour flight. The spherical metal emblems inside the probe signify that it was initially intended to crash into the Moon. By missing its target, however, it became the first artificial object in solar orbit, and still circles to this day between Earth and Mars. The scientific instruments on board made it possible to measure the low density of high-energy particles in the outer Van Allen belt, as well as the absence of a lunar magnetic field and the presence of the solar wind.

The R-7 rocket, based on the first Soviet intercontinental missile, launched the Sputnik satellites as well as the Luna probes. Its fuel was a mixture of kerosene and liquid oxygen.

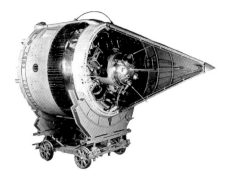

The Luna 1 probe under its protective nose cone, in launch configuration.

OPPOSITE PAGE, TOP
Luna 1 replica attached to the third stage of the R-7 rocket. Following its flyby of the Moon, the Luna 1 probe became the first artificial object placed in heliocentric orbit.

OPPOSITE PAGE, BOTTOM
Replica of Luna 2, the first artificial object to touch the Moon. The spherical structure had several antennas and contained scientific instruments, the electrical supply and several emblems with the Soviet coat of arms to mark the event.

With the Luna probes, the dream of conquering space became a reality

Luna 2 was successfully launched by an R-7 8K72 rocket on September 12, 1959, with the goal of touching the moon for the first time. This second probe of the Soviets' lunar program weighed 390 kilograms, was identical in structure to Luna 1, and likewise did not have its own independent propulsion system. In order to facilitate tracking the spacecraft from Earth and to continue to study the behavior of a gas in interplanetary space, Luna 2 was also programmed to discharge a cloud of sodium gas. In addition to Luna 1's instruments, Luna 2 carried an ingenious ion trap conceived by Konstantin Gringauz, which allowed it to demonstrate without question the existence of the solar wind. It also confirmed the absence of radiation belts around the Moon. Luna 2 crashed into the Moon in the Sea of Rains on September 13, 1959, after slightly more than 33 hours of flight, and was followed 30 minutes later by the rocket's third stage. As with Luna 1, Luna 2 carried emblems with Soviet coat of arms, which were to be disseminated on the Moon's surface. Luna 2's speed of impact was estimated as more than 3 kilometers per second, which means that both the probe and the third stage were probably vaporized upon impact. ☾

TRAJECTORY OF LUNA 2

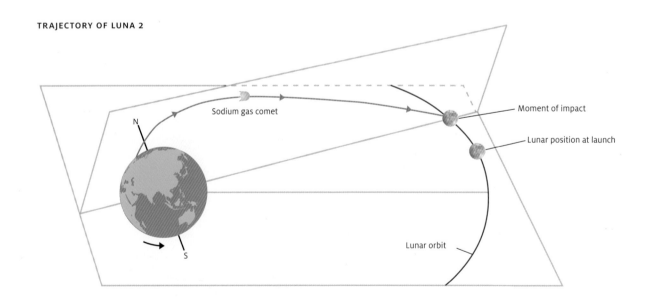

Sodium gas comet

Moment of impact

Lunar position at launch

Lunar orbit

N

S

1959   Pioneer 4

NASA technicians in sterile suits inspect the Pioneer 3 probe prior to its shipment to the launch center at Cape Canaveral. A problem with the Juno II propulsion rocket caused it to crash back to Earth.

OPPOSITE LEFT

The Juno II launcher of both the Pioneer 3 and 4 missions was a modified civilian version of the Jupiter medium-range nuclear ballistic missile.

OPPOSITE RIGHT

Pioneer 4 carried but a small payload of about 6 kilograms.

PIONEER 4, launched by a Juno II rocket on March 3, 1959, from Cape Canaveral in Florida, was the first American probe to free itself from Earth's gravitational pull. Its mission was to approach the lunar surface to a distance of 60,000 kilometers on March 4, 1959, and then enter solar orbit. The scientific equipment, weighing but a few kilograms, consisted of a Geiger-Muller-type radiation counter and a photographic device, which could not be used. The probe did not detect any radiation emanating from the Moon. ☾

Luna 3

T HE LUNA 3 PROBE was the first spacecraft equipped with solar panels. Weighing 279 kilograms, this third Soviet lunar probe was more sophisticated than its predecessors. Among other features, it was equipped with a new attitude-control system, an improved electrical system and a complex system to transmit photographic images. In fact, it was designed specifically for lunar photography.

Launched from Baikonur on October 4, 1959, on an R-7 8K72 rocket, its mission consisted of photographing the far side of the Moon for the very first time. Having skimmed past the Moon at a distance of 6,000 kilometers, the space probe went around our natural satellite, following a trajectory that eventually brought it back closer to Earth. On October 7, 1959, Luna 3 took its place in history. During 40 minutes, from an altitude of about 60,000 kilometers, the Soviet probe took 29 photographs, covering 70 percent of the Moon's hidden face. The activation of the historic series of photographs was initiated by a photoelectric cell sensitive to moonlight.

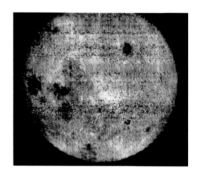

The first photograph of the Moon's far side, transmitted by Luna 3 in 1959.

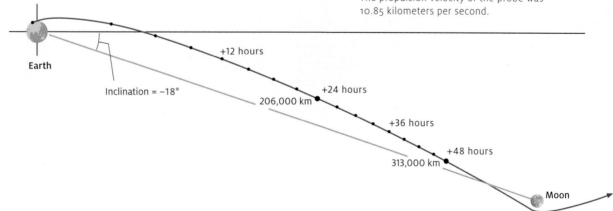

TRAJECTORY OF LUNA 3
The propulsion velocity of the probe was 10.85 kilometers per second.

Earth

Inclination = −18°

+12 hours

206,000 km

+24 hours

+36 hours

+48 hours

313,000 km

Moon

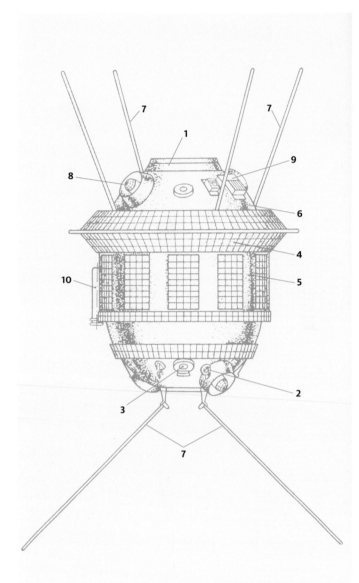

Luna 3 is the first space probe equipped with solar panels.

RIGHT

Yenisey-2, the remarkably compact photo-television system aboard Luna 3.

MAIN ELEMENTS OF LUNA 3

1. Camera compartment (closed cover)
2. Attitude-control thrusters
3. Solar tracker
4. Solar panels
5. Solar panels
6. Hermetically sealed box
7. Radio antennas
8. Ion trap
9. Micrometeorite counter
10. Mass spectrometer

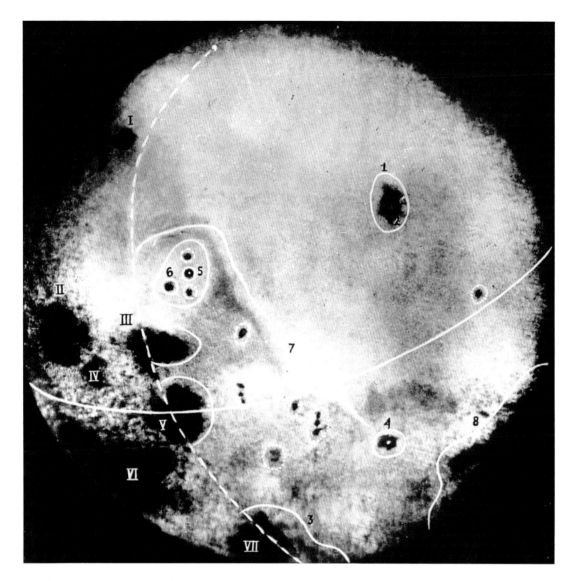

The imaging system of Luna 3, christened the Yenisey-2, consisted of a twin-lens 35 mm camera, a 200 mm f/5.6 lens and a 500 mm f/9.5 lens. The 200 mm lens could capture images of the entire lunar disc, while the 500 mm could capture surface details of a particular region. The film unwinding mechanism was coupled with an optical calibrating system with a photo-multiplier. An image was scanned at a resolution of 1,000 horizontal lines. Though very grainy, 17 usable images were transmitted by radio communication during the return voyage of the probe toward Earth. All contact with Luna 3 was lost on October 22, 1959. ☾

The first composite map of the far side of the Moon, as photographed by Luna 3, showing the Moscow Sea (1), the craters Tsiolkovsky (4), Lomonosov (5) and Joliot-Curie (6), as well as the Sea of Fertility (VI). The solid line indicates the equator and the dotted line the border between the visible and the far side of the Moon.

1961   The Ranger Program

The first image of the lunar surface transmitted by an American spacecraft.

RIGHT

The Jet Propulsion Laboratory Control Center for robotic missions as it appeared in 1964.

OPPOSITE PAGE

The Ranger 6 spacecraft.

T HE U.S. RANGER PROGRAM, first developed in 1959, had as its goal the production of the first close-up images of the lunar surface, in anticipation of a lunar landing. The Ranger probes were designed to take a series of photographs of the lunar surface until the very moment of impact. The sequential failures of Block I (Ranger 1 and 2) and Block II (Ranger 3, 4 and 5) were an embarrassment to NASA, especially in light of the race to the Moon with the Soviets. They had to wait until Block III (Ranger 6, 7, 8 and 9) to reach this goal. Among the many technological innovations of the Ranger probes were the independent propulsion system to secure a better trajectory precision and three axes-attitude-stabilization systems, which enabled the probe to have a fixed orientation. This stabilization capacity had the advantage of being able to constantly orient the solar panels toward the Sun and the radio antenna toward Earth,

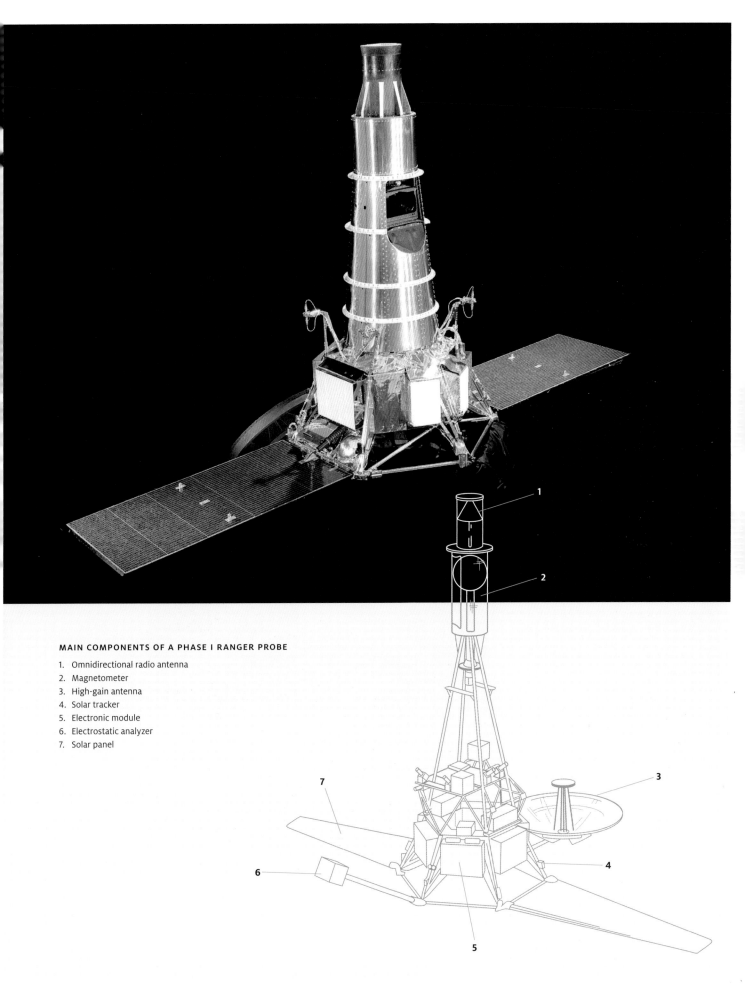

MAIN COMPONENTS OF A PHASE I RANGER PROBE

1. Omnidirectional radio antenna
2. Magnetometer
3. High-gain antenna
4. Solar tracker
5. Electronic module
6. Electrostatic analyzer
7. Solar panel

Points of impact for the various Ranger probes.

while keeping the cameras aimed at their targets. The Block III Ranger probes were equipped with six television cameras (four telephoto lenses and two wide-angle lenses) capable of shooting at a rate of 300 images per minute. Launched on January 30, 1964, on board an Atlas-Agena B rocket, Ranger 6's trajectory was perfect, but its television system did not go on as planned. Launched on July 31, the Ranger 7 mission was a total success. Ranger 7 transmitted 4,316 images before crashing into a region between Mare Cognitum and Oceanus Procellarum. Ranger 8 followed on February 17, 1965, and after transmitting 7,137 quality images of the lunar surface, it crashed on February 20 in the region of the Mare Tranquillitatis. The last transmitted image had a resolution of 1.5 meters. Ranger 9, launched March 21, 1965, ended this observational program with 5,814 high-quality images of the crater Alphonsus. The high resolution of these final images revealed objects as small as 30 centimeters before the probe made impact at 2.67 kilometers per second. This level of resolution was 1,000 times better than was possible with Earth-bound telescopes. These new images revealed numerous impact craters of all sizes, which have characterized the surface of the Moon since its formation. ☾

The six television cameras of Ranger 6, 7, 8 and 9 were programmed to take bursts of images, continuing until their impact on the lunar soil.

The Zond Program

THE ZOND (Russian for "probe") series of spacecraft were the first Soviet vessels destined for interplanetary voyages. They were intended for use in the exploration of Venus and Mars, as well as for crewed missions to circle the Moon. The Venus probe, Zond 1, which launched on April 8, 1964, could not accomplish its objectives due to a radio communication problem during launch. Zond 2 launched on November 30, 1964, intended for Mars, but it was lost due to interruption in radio contact in May of 1965. Zond 3 was prepared to accompany Zond 2. Having missed the launch window for Mars, Zond 3 program managers decided to cut their losses with a flyby of the Moon's far side, to obtain scientific data and test their system of long-distance communication. On July 18, 1965, Zond 3 took off from Baikonur, propelled on an interplanetary trajectory by an R-7 rocket. Weighing 960 kilograms, the imposing spacecraft was equipped with an array of scientific instruments. These included ultraviolet and infrared spectrographs, a magnetometer, radiation detectors, micrometeorite counters, an experimental ion engine and a camera with a 106 mm lens. An automatic film-development mechanism was coupled with a television system for the transmission of images via radio. On July 20, 1965, Zond 3 flew over the far side of the Moon at an altitude of 9,200 kilometers. The probe took 25 excellent-quality images of the lunar surface in a span of 68 minutes, which it then transmitted over a distance of 2,200,000 kilometers and again from a distance of 31,500,000 kilometers. Radio signals continued to be received from as far away as Mars' orbit, confirming the Soviet's capability to communicate over very long distances.

Transport of the powerful Proton rocket. This type of rocket was used to send Zond vessels into circumlunar trajectories.

Replica of a Zond spacecraft.

One of the first Zond 3 images of the Moon.

Zond 7 was able to send color images of the Moon and the Earth.

The Zond missions that followed (from Zond 4 in 1968 to Zond 8 in 1970) served mainly to test trajectories for circumlunar trips and the return of the capsules to Earth, in preparation for crewed missions. ☾

The photo-television camera on board Zond 3 was a marvel of miniaturization at the time.

RIGHT
The re-entry capsules from Zond 5 (left) and Soyuz 3 (right) are on display at the Energia Museum in Moscow.

OPPOSITE PAGE
Earthrise captured by Zond 7 in 1969.

1966

Luna 9

BELOW

The Lunar 9 vessel and lander.

OPPOSITE PAGE

Replica of the Lunar lander, shown in operational mode with its petals deployed.

PROPELLED BY A MOLNIYA (Russian for "lightning") rocket, a four-stage version of the R-7, Luna 9 was launched toward the Moon on January 31, 1966, from the Baikonur Cosmodrome. On February 3, 1966, Luna 9 became the first spacecraft in history to soft land on another celestial body. On making contact with the surface, the landing module activated the ejection mechanism of the 99-kilogram, sphere-shaped payload. Protected by an external inflatable bag, Luna 9 stabilized itself on the ground after opening its four petals, and became operational. The images transmitted from Lunar 9 were the first from an extraterrestrial surface. In the morning of February 4, the camera equipped with a movable mirror produced a panoramic view of the landing site, situated in the Oceanus Procellarum, from nearby rocks to the horizon 1.4 kilometers away. The lunar probe continued to transmit photographic images and data until the batteries died three days later. The successful Luna 9 mission was a source of pride for the Soviet space program, as this was its twelfth attempt at a Moon landing. The mission also produced information of the utmost importance: that it was now possible to land on the Moon without sinking into the dust. ☾

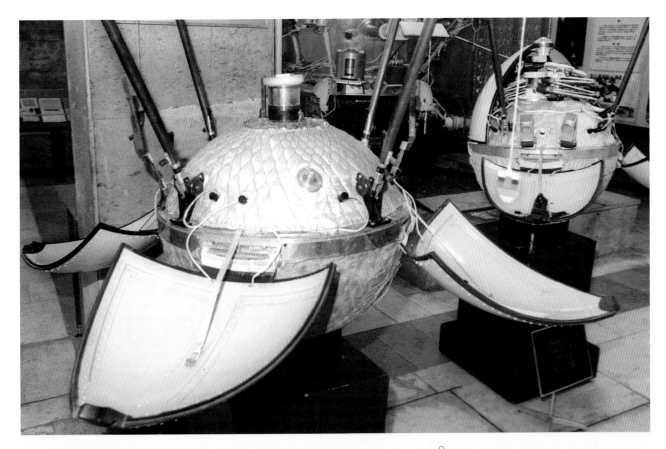

MAIN COMPONENTS OF LUNA 9

1. Panoramic camera
2. Mirror
3. Antenna
4. Optical calibration target
5. Radiation detector
6. Stabilizing petal

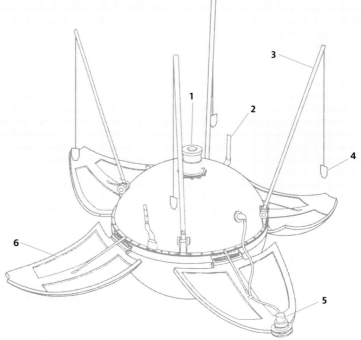

Panoramic view as initially photographed by Luna 9 from a polar perspective, and the same panoramic view after reconstruction in a horizontal plane, below.

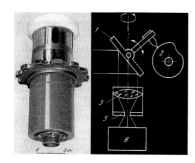

The miniature panoramic camera on Luna 9 consisted of an ingenious opto-mechanical system that covered the entire scene by sweeping motions.

First image transmitted from the site of the Luna 9 landing.

Luna 9 Scoop!

It so happened that the radio communications from Luna 9 were compatible with the standard international protocol then used by journalists to transmit images via fax machines. The radio signals from Luna 9 that were captured by the powerful Jodrell Bank radio telescope in England were then converted into images with the help of the fax machine on loan for the occasion by the *Daily Express* newspaper. Since the Soviet authorities had delayed the transmission of images from Luna 9, the first lunar landing in history received rapid and international media coverage before its official announcement. This became an embarrassing situation for those in charge of the Soviet space program, who subsequently tried their best to jam the radio signals in order to avoid another Western scoop.

1966 Luna 10

A commemorative Soviet stamp of the Luna 10 mission. Placing a probe in orbit around another celestial object was a historic first.

OPPOSITE PAGE
Luna 10 was placed into orbit around the Moon on April 3, 1966.

O N APRIL 3 1966, the Luna 10 probe became the first artificial satellite to orbit the Moon and de facto, the first satellite to orbit another celestial body. Launched March 31, 1966, by a Molniya rocket, Luna 10 carried with it an array of scientific measuring instruments including, among others, a gamma-ray spectrometer, a magnetometer and a micrometeorite collector, but no camera. Three days later, the 245-kilogram satellite entered into lunar orbit, approaching to within 350 kilometers of the surface every 2 hours and 58 minutes. The deviations within Luna 10's orbit made it possible to detect, for the first time, gravitational anomalies due to the mass concentrations (mascons) within lunar basins. The presence of these mascons was confirmed by the Lunar Orbiter 5 mission (see page 50) a year later. ☾

Luna 10 plays "The Internationale"

The Luna 10 probe was programmed to directly transmit the music of "The Internationale" during the 23rd Congress of the Communist Party. The rehearsal went off as planned, but a technical problem on the day of the event, April 4th, 1966, forced engineers to use a prerecorded magnetic tape and pretend that it was actually a direct transmission from the Moon.

1966   The Surveyor Program

The successful night launch of the Atlas-Centaur rocket for the Surveyor 6 mission, in November 1967.

T HE NASA SURVEYOR PROGRAM experimented for the first time with robotic landing procedures, in preparation for the future crewed Apollo missions. The main goals of the program were to study the resistance characteristics of the lunar soil and to document in images the geologic environment of the landing sites. Surveyor 1, built by Hughes Aircraft, launched on May 30, 1966, aboard an Atlas-Centaur rocket at Cape Canaveral. It landed successfully on the Moon on June 2, 1966, four months after Luna 9, near the crater Flamsteed, located southwest of Oceanus Procellarum. In spite of the time lag, this first American lander reassured NASA of its ability to remain in the race with the Soviets. The launch trajectory was aimed at a direct impact on the Moon, cushioned by retrorockets during the final phase of approach. The retrorockets were accurately programmed to stop the lander 3 meters above the lunar surface before releasing it. The Moon's surface proved to be solid enough to sustain the impact of a 300-kilogram machine. During the 65 hours planned for this mission, Surveyor's television camera, with its two modes of resolution, with 200 to 600 scan lines, captured 11,237 images of the site, including the very first color images. Surveyor 2, launched on September 20 of the same year, was not quite so lucky. It crashed into the Moon because of a technical problem caused when a vernier motor malfunctioned during a trajectory correction. Surveyor 3, launched April 17, 1967, landed April 20 at a site on Mare Cognitum in Oceanus Procellarum, where it transmitted 6,315 images.

The landing of Surveyor 3 was turbulent. The descent radar, fooled by the reflections from some rocks, did not send the command to stop the retro-rockets at the right moment. The probe bounced 10 meters above the ground, followed by another jump of 3 meters before setting down. This mission was the first to carry a shovel attached to the end of a telescoping arm to dig trenches and collect samples to study the mechanical properties of the lunar soil. The landing site of Surveyor 3 was visited in 1969 by the Apollo 12 mission, when the lunar module Intrepid landed less than 200 meters from it. Astronauts Alan L. Bean and Pete Conrad were able to return several pieces of the probe to Earth for analysis, including the television camera. On July 17, 1967, Surveyor 4 was getting ready for its landing when telemetry contact was lost, probably caused by an explosion. Surveyor 5 was the third successful American lunar landing craft, touching down on September 11, 1967. That mission was a complete success and the probe transmitted more than 19,000 images from Mare Tranquillitatis. Its articulated arm

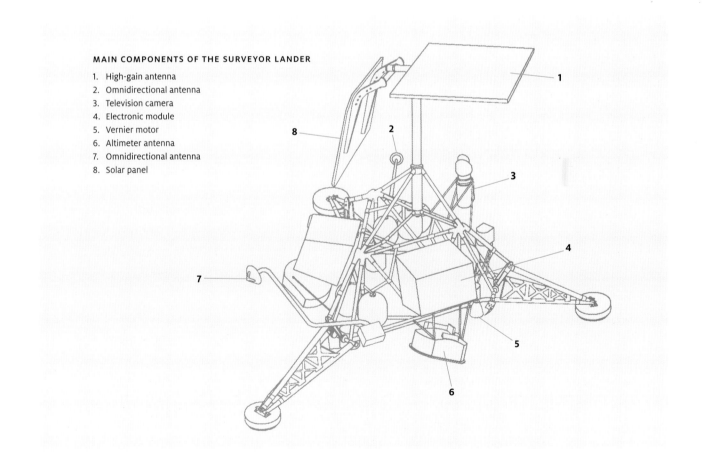

MAIN COMPONENTS OF THE SURVEYOR LANDER

1. High-gain antenna
2. Omnidirectional antenna
3. Television camera
4. Electronic module
5. Vernier motor
6. Altimeter antenna
7. Omnidirectional antenna
8. Solar panel

was equipped with an X-ray spectrometer akin to the one in the Martian rover, Sojourner (see page 166), used to analyze, for the very first time, the composition of the chemical elements of lunar soil. The results revealed the presence of basaltic rocks. Surveyor 6, which landed on November 7, 1967, was similar to Surveyor 5 in all aspects and transmitted more than 30,000 images of the Sinus Medii area. Surveyor 6's rockets ignited again on November 17, and it rose up to 4 meters and landed a short distance farther away. This test did not in any way affect its functioning. The last mission of this series programmed Surveyor 7 to land near Tycho, an immense crater visible with the naked eye from Earth. It landed on January 10, 1968, less than three kilometers from its target. The probe transmitted more than 21,000 images but the X-ray spectrometer could not be deployed correctly, and radio communications became sporadic after battery power loss during the first night. By the end of the Surveyor program, NASA had gained enormous expertise in lunar landings. ☽

After landing, Surveyor 6's retrorockets were fired again to make the lander jump and test the mechanical resistance of the soil, in preparation for crewed landings.

Thanks to a new radar system that facilitates accurate lunar landings, the Apollo 12 mission would pay a visit to Surveyor 3 on November 19, 1969. Astronaut Alan Bean retrieved several pieces of the probe in order to analyze the effects of a prolonged stay (two years) on the Moon. The lunar module is evident on the horizon.

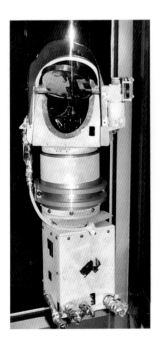

Ultra-resistant bacteria?

The inspection of the Surveyor 3 camera revealed the presence of the common bacteria *Streptococcus mitis*. At the time, NASA concluded that the probe had not been sterilized adequately prior to liftoff, and the bacteria had survived the extreme conditions of the lunar environment. However, the actual interpretation favors a more simple explanation: several breaches in the sterile procedures took place during return of the capsule on the Apollo 12 mission and during the microbiological analysis on Earth. As a result of this incident, NASA submits all its space probes to draconian sterilization procedures, so as to avoid all possible risk of contamination of other planets by terrestrial microbes.

The Lunar Orbiter Program

The launch of Lunar Orbiter 1 inaugurated a fruitful series of missions between August 1966 and August 1967, which resulted in complete photographic coverage of the Moon of such high quality that it remains unequaled to this day.

OPPOSITE PAGE

Main Components of Lunar Orbiters.

THE GOAL OF NASA'S LUNAR ORBITER PROGRAM was to send a series of probes into low orbit (40 to 1,850 kilometers) around the Moon in order to map with great cartographic precision the 20 possible landing sites for the future Apollo missions. The five Lunar Orbiters launched between August 10, 1966, and August 1, 1967, clearly attained their objectives. The on-board cameras had two lenses, a 610 mm telephoto and an 80 mm wide-angle. The 70 mm film was chemically developed on board, dried and then digitized before being transmitted back to Earth via a high-gain antenna. The Lunar Orbiter program provided 2,180 high-resolution photographs for study and 882 lower-resolution images, covering 99 percent of the lunar surface. Lunar Orbiter 1 took the first photographs of the Earth seen from the Moon, including the first earth-rise. Some images taken by Lunar Orbiter 3 were so detailed (with resolution down to 1 meter) that the Surveyor 1 lander was identified on the lunar surface. Lunar Orbiter 5, launched August 1, 1967, was able to confirm the gravitational anomalies, termed mascons, detected by the Soviet probe, Luna 10 (see page 44). The simultaneous presence of several orbiters around the Moon greatly facilitated the Americans' space communication capabilities. Ultimately all lunar orbiters were crash-landed unto the lunar surface, so as not to interfere with Apollo missions. ☾

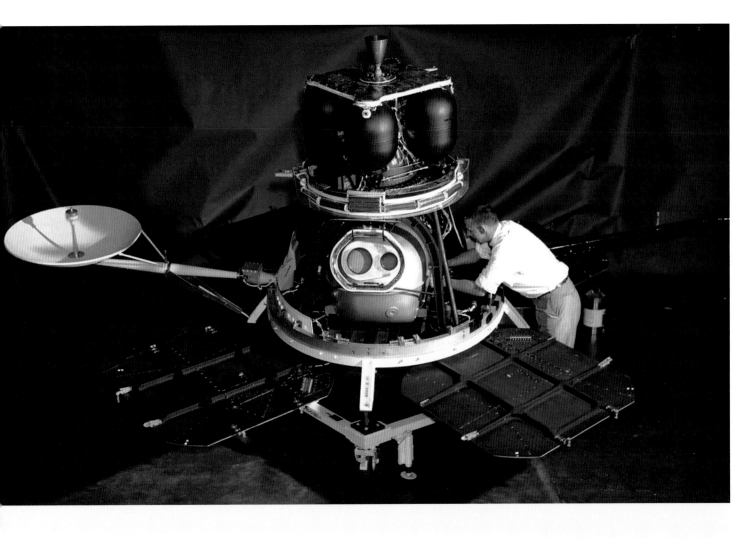

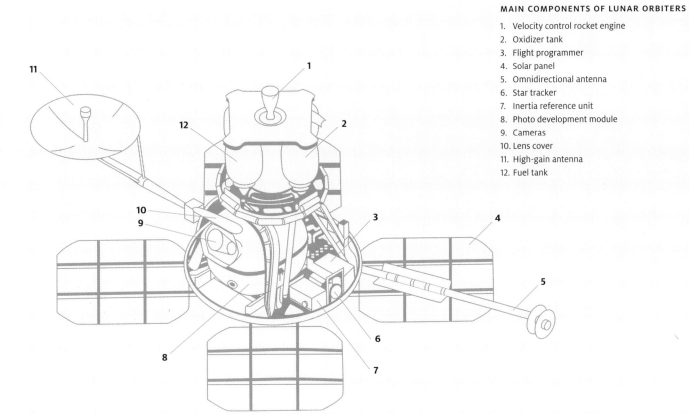

MAIN COMPONENTS OF LUNAR ORBITERS

1. Velocity control rocket engine
2. Oxidizer tank
3. Flight programmer
4. Solar panel
5. Omnidirectional antenna
6. Star tracker
7. Inertia reference unit
8. Photo development module
9. Cameras
10. Lens cover
11. High-gain antenna
12. Fuel tank

Lunar Orbiter's cameras provided high-resolution images of the lunar surface.

RIGHT

The first photograph transmitted by Lunar Orbiter 1 of the Earth as seen from the Moon. This image was subsequently reworked with modern digital technology.

OPPOSITE PAGE, TOP

Oblique view of the crater Copernicus, captured by Lunar Obiter 2.

OPPOSITE PAGE, BOTTOM

Close-up of Copernicus, showing details of a mountainous chain in the center of the crater, 96 kilometers in diameter and 3 kilometers deep. This was to be the destination of the Apollo 20 mission, which was cancelled. This picture was deemed "Picture of the Century" by the news media at the time.

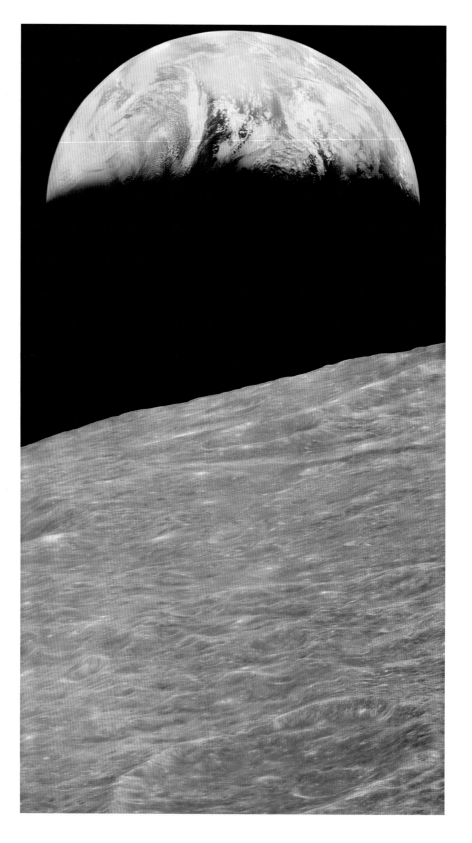

1966

Luna 11, 12 and 14

Aᴀᴛᴇʀ ᴘᴏsɪᴛɪᴏɴɪɴɢ ɪɴ ᴀ ᴛʀᴀɴsғᴇʀ ᴏʀʙɪᴛ around the Earth on August 24, 1966, the Luna 11 probe was placed into lunar orbit on August 28. This orbit had an inclination of only 27° (compared to 72° for the Luna 10 probe), in order to focus on the equatorial regions of the Moon and locate potential landing sites for future crewed missions. The mission's principal objectives were twofold: a study of the chemical composition of the Moon by measuring X-ray and gamma-ray emissions and a search for gravitational anomalies. No imaging system was on board Luna 11, unlike its twin Luna 12, which was equipped with a television camera similar to those on the Zond probes, to provide the first close-up views of the lunar surface from orbit. Luna 12 was launched on October 22, and after a transitional Earth orbit, placed into lunar orbit on the 25ᵗʰ. Its cameras could observe areas of 25 square kilometers, with resolution capable of detecting craters 20 meters in diameter. The Luna 12 mission was considered a technical success, despite the fact that many of its photographs were never seen and very few of those images were made public. Contact with Luna 12 was lost on January 19, 1967, after 602 orbits and depleted battery power. Since that time, this probe probably crashed into the moon, since the gravitational pull of the mascons progressively deform the orbital trajectories. Very little information was provided about Luna 14 after its launch on April 7, 1968, other than that it was similar in structure to Luna 12. ☾

One of the few available images of the lunar surface taken by Luna 11 and Luna 12.

OPPOSITE PAGE
Luna 12 orbital module.

1966

Luna 13

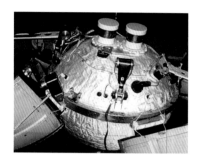

Close-up view of the Lunar 13 lander in operational configuration.

BELOW AND OPPOSITE PAGE
Panoramic images of the lunar surface transmitted by Luna 13.

THE YEAR 1966 was unusually productive at the Soviet Cosmodrome in Baikonur. The year ended with the launch of Luna 13 on December 21, atop a Molniya rocket. Luna 13 was positioned into Earth orbit before beginning its translunar trajectory and landing on December 24, 1966. Luna 13 thus became the third lunar lander in history, after Luna 9 and Surveyor 1. A few minutes after its contact with the lunar surface, northwest of Oceanus Procellarum, Luna 13 began radio communications and provided information about its environment. Several panoramic photographs of the site were taken from different angles of lunar illumination. Each panoramic image took 100 minutes to transmit. A second camera was to provide stereoscopic data, but it remained non-functioning. The five very detailed panoramic views, revealed a terrain less hilly than that of the Luna 9 site. Luna 13 was also equipped with a penetrometer to measure the force required to penetrate the lunar regolith. Having successfully accomplished its mission, the probe died on December 28, 1966, due to battery depletion. ☾

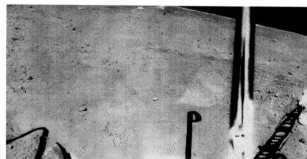

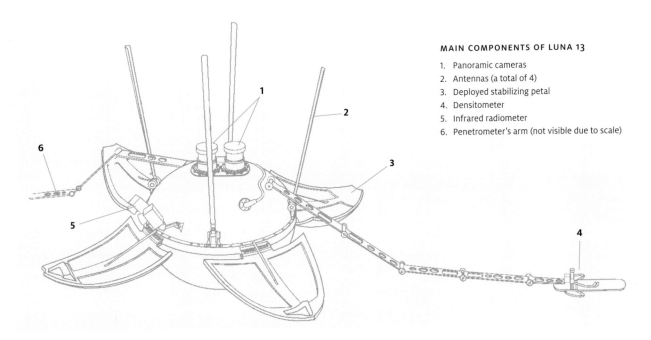

MAIN COMPONENTS OF LUNA 13

1. Panoramic cameras
2. Antennas (a total of 4)
3. Deployed stabilizing petal
4. Densitometer
5. Infrared radiometer
6. Penetrometer's arm (not visible due to scale)

Luna 16, 20 and 24

The Proton rocket, which in the 1970s launched Luna 16, Luna 20 and Luna 24 into Earth–Moon trajectories, is still in use today, making it one of the most reliable space vehicles ever built.

OPPOSITE PAGE
Final assembly of the Luna 16 return module.

The soviet robotic mission Luna 15 was launched on July 13, 1969, and designed to bring the very first lunar soil samples back to Earth. Today it is interpreted as an unfortunate attempt to garner world attention and counter the success of the American Apollo program. In fact, the Eagle lunar module of the Apollo 11 mission was to land on the Moon on July 20, 1969, and the two astronauts on board were to walk on the lunar soil. Luna 15 crash-landed during its final descent maneuvers. The Soviets had definitely lost the race to the Moon. Nevertheless, Luna 15 marked an important milestone in international collaboration: its flight plans were revealed to avoid a collision with Apollo 11.

A similar sample-return mission, Luna 16, was launched on September 12, 1970, from the Cosmodrome at Baikonur. The massive robotic spacecraft, weighing more than five tons, required a powerful Proton K rocket (this type of rocket is still being used). The landing of Luna 16 took place on September 20 in the Sea of Fertility, to the east of the Webb crater. The sampling system consisted of an articulated arm equipped with a digging tool that could penetrate up to 35 centimeters into the lunar soil. A semicircular motion of the arm was used to transfer the regolith sample into a spherical capsule located on top of the return module. On September 21, the upper stage of Luna 16 ignited its engines and took off toward Earth. Three days later, on September 24, the capsule, with its precious 105 grams of lunar soil, opened its descent parachute and landed in Kazakhstan. Mineralogical analysis of the sample confirmed its basaltic origin, nearly identical to that returned by the Apollo 12 team.

Despite its technical complexity, the ambitious Luna 16 expedition was a total success. The parties responsible for the Soviet space program greatly needed some form of success during this period of American domination. The Luna 20 mission was launched on February 14, repeated the same landing procedures and returned with 55 grams of lunar soil on February 25. This time, the soil sample taken in the plateau region, 160 kilometers from the Luna 16 site, was mainly composed of anorthosite rather than basalt.

The Luna 24 mission, which lifted off from Baikonur on August 9, 1976, turned out be the last Soviet lunar mission. In fact, it constituted a final attempt to collect samples of regolith in Mare Crisium, since the preceding Luna 23 mission was aborted, due to a technical malfunction during liftoff.

BELOW
Verification of Luna 16's telecommunication system

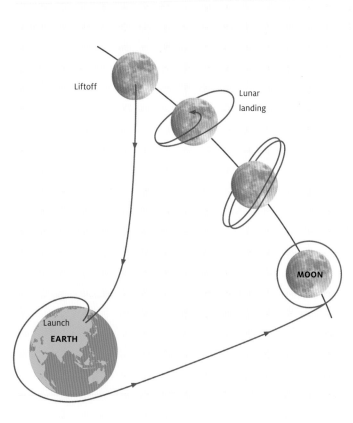

TRAJECTORY OF THE LUNA 16 MISSION

Liftoff

Lunar landing

MOON

Launch
EARTH

Luna 24 differed from the other "two-way" missions mainly by its method of soil collection. Luna 24 could retrieve samples from a depth of 2.5 meters, thanks to a collecting system on rails. Fragments of soil were taken and kept in a cylindrical tube that was rolled up in a spiral fashion in a hermetically sealed container. On August 22, 1976, the Luna 24 return capsule, with its 170 grams of soil from Mare Crisium, landed intact in the Surgut region of Siberia. The analysis of minerals taken at different depths on the Moon helped to clarify the origins and formation of regolith. Like the soil samples brought back by Luna 16 and Luna 20, those of Luna 24 were shared with scientists around the world in the spirit of collaboration. Luna 24 remains, to this day, the last lunar lander from any country. ☽

The return capsule of Luna 20, shortly after its landing on a plain in Kazakhstan on February 25, 1972.

MAIN COMPONENTS OF LUNA 16

1. Lunar sample collecting capsule
2. Omnidirectional antenna
3. Return to Earth propulsion rocket
4. Low-gain antenna (a total of 4)
5. Attitude-control thruster
6. Propellant tank
7. Sample collector arm
8. Digging head of the collector in hermetically sealed container

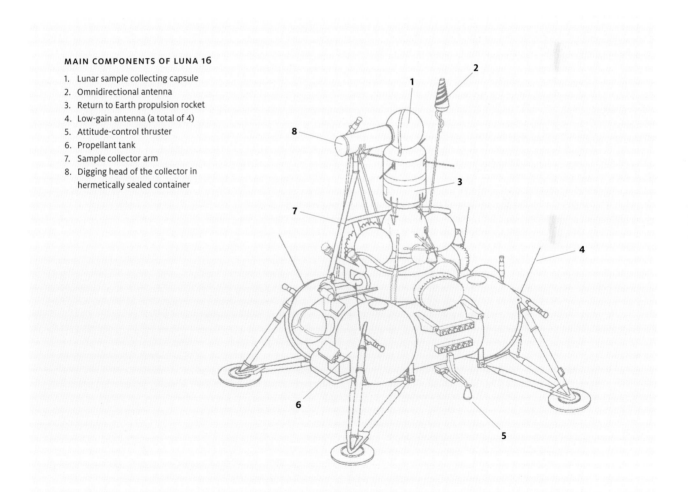

1970

Luna 17 and 21

ANOTHER AMBITIOUS SOVIET MISSION, Luna 17, took off on November 10, 1970, from Baikonur. This time it involved landing a motorized robot, or rover, remotely guided from Earth. Launched by a Proton K rocket, Luna 17 went into a stationary Earth orbit before transferring into a lunar orbital trajectory on November 15. Landing occurred without incident on November 17 in Mare Imbrium, a huge basaltic plain. A few hours later, the Lunokhod 1 rover descended from its mothership and set its eight independently mobile wheels onto the lunar soil. Equipped with numerous mobile cameras to document its road trip, the rover was assigned the task of analyzing the soil with a penetrometer and an X-ray spectrometer. It had an estimated operational lifespan of three lunar days, but lasted for 11 lunar days (322 Earth days). During this time, it covered more than 10 kilometers and transmitted more than 20,000 televised images and several high-resolution panoramic photographs. Capable of moving at a maximum speed of 100 meters per hour, it was piloted by a five-man Earth-based crew, who had to take into account the

RIGHT

Overview of the Luna 17 lander with the Lunokhod 1 Rover and exit ramps deployed.

OPPOSITE PAGE

Model of the Lunokhod 1 rover in mobile configuration, with its solar panel and laser reflectors open.

MAIN COMPONENTS OF THE LUNOKHOD

1 ROVER

1. Omnidirectional antenna
2. High-gain directional antenna
3. Solar panel (not deployed)
4. Panoramic camera (a total of 4)
5. X-ray spectrometer
6. Television cameras
7. Laser reflector

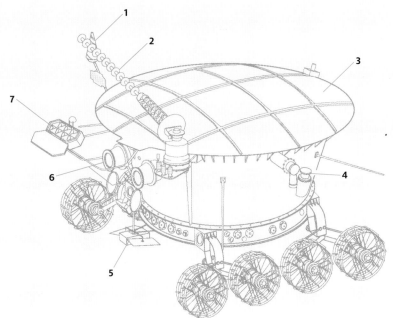

The Lunokhod differed from its predecessor primarily by having three television cameras and a traction system with eight more efficient wheels.

BELOW
Panoramic mosaic of photographs taken by Lunokhod 2 of its landing site.

five-second communications delay due to the distance between the Earth and the Moon. Along with the successes of Luna 16 and Zond 8, Luna 17 and its Lunokhod 1 made 1970 a glorious year for the Soviet engineers.

Confident, the Soviets decided to send a second generation of the rover to the Moon a few years later. The Luna 21 mission, with its Lunokhod 2 on board, took off on January 8, 1973, less than a month after the last Apollo mission, and landed on the Moon, as planned, on January 15, 1973, in the crater Le Monnier. The Lunokhod 2 rover benefited from several improvements over the Lunokhod 1, including a more modern set of television cameras and an improved eight-wheel traction system. More mobile than its predecessor, it ultimately covered more than 37 kilometers, transmitted more than 80,000 television images and photographed more than 80 lunar panoramas. Lunokhod 2's laser reflector, built by a French team, allowed them to measure the Earth–Moon distance within a precision of 20 centimeters. On May 9, 1973, Lunokhod 2 accidentally fell into a small crater, from which it could not extricate itself. Dust accumulation on its solar panels and radiator affected its internal temperature regulation, which plunged irreversibly low. On June 3, Soviet mission control announced loss of radio contact and hence the end of the mission.

Several more rover projects were planned, but it has been more than 30 years since the lunar surface was last disturbed by Lunokhod 2. ☾

Luna 19 and 22

THE LUNA 19 AND 22 ORBITERS were massive and sophisticated observation crafts, designed to undertake prolonged scientific analyses of the lunar environment. They were built on the same framework as that of Luna 17, except that the rover Lunokhod was replaced by a pressurized compartment containing the science instrumentation.

Luna 19 left Earth September 28, 1971, on a Proton K rocket and inserted itself into a circular operational lunar orbit on October 1, 1971, at an altitude of 140 kilometers. Its twin, Luna 22, lifted off on May 29, 1974, and moved into a quasi-circular (219 by 222 kilometers) lunar orbit on June 2 of that year. In addition to photographic coverage of the flyover areas, thanks to the several on-board cameras (including a panoramic one), Luna 19 and Luna 22 sent back data on the gravitational anomalies (mascons, see page 44), the lunar magnetic field and the mineral composition of the surface. The Luna 19 mission did not end until October 1972, after more than 4,000 orbits. The Luna 22 mission, which lasted 15 months until November 1975, was also a total success. It remains, to this day, the last lunar orbital mission of the Soviet space program. ☾

Model of the Luna 19 orbiter.

BELOW
Panorama of the lunar globe, compiled from Luna 22 data.

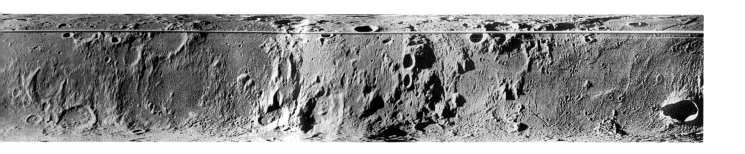

1990 Hiten-Hagoromo

The space probe Hiten with its small passenger, Hagoromo, on top.

THE GOAL of the Japanese Hiten mission was to place a satellite into orbit around the Earth to test technologies intended for lunar orbital missions. The Hiten ("musical angel") satellite carried with it the small Hagoromo (the name of a legendary feather coat), which was to be launched in the neighborhood of the Moon. Launched by an M-3S rocket on January 24, 1990, Hiten assumed a highly elliptical geocentric trajectory that would circle the Moon with every orbit. While the launch of Hagaromo went as expected, its radio transmitter failed, making it impossible to determine if it had entered the planned lunar orbit. The Japanese controllers then proposed to salvage the mission by placing the entire Hiten satellite into lunar orbit, since the spacecraft had only used 10 percent of its fuel at that point. American mathematician Edward Belbruno, an expert in orbital dynamics, calculated a trajectory that would require very little energy. On October 2, 1991, Hiten attained the orbit calculated by Belbruno, bypassing the Moon and reaching the Lagrangian points L_4 and L_5 of the Earth–Moon system. Hiten was equipped with only one scientific instrument, a micrometeorite counter, which did not detect any at the Lagrangian points. In order to adjust its trajectory, Hiten used the technique of aerobraking in the Earth's atmosphere, a navigational method that had not yet been tried at that point but is now used routinely. On February 15, 1993, during a pass of just 422 kilometers above the Moon's surface, Hiten utilized its remaining fuel to move into stable lunar orbit. After two months, the last bit of fuel was used to crash-land Hiten into the Moon on April 10, 1993, south of the crater Petavius. For the first time since the 1960s, when the Soviets and Americans were primarily involved, a third nation, Japan, had developed a space program capable of sending a probe to the Moon. ☾

Clementine

T HE SMALL PROBE Clementine marked the first U.S. return to the Moon after the Apollo program. The product of a cooperative effort between NASA and Strategic Defense Initiative Organization (SDIO), derived from the 1977 Ballistic Missile Defense Organization, Clementine was part of the U.S. Department of Defense's "Star Wars" initiative under the Reagan administration. The military part of the mission was testing the performance of small launch vehicles and the performance of various light material composites under the extreme temperature conditions of interplanetary space. The scientific part of the mission was twofold: to study various aspects of lunar geology (selenology) and to rendezvous with the asteroid Geographos. Clementine was launched from Vandenberg Air Force Base in California on January 24, 1994, on board a Titan 23G rocket. This rocket was one of 14 nuclear missiles converted to civilian launchers.

BELOW, LEFT
Final inspection of the Clementine probe before launch.

BELOW
Artist's rendition of Clementine in operation, with solar panels extended.

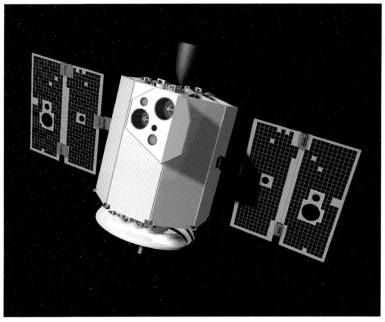

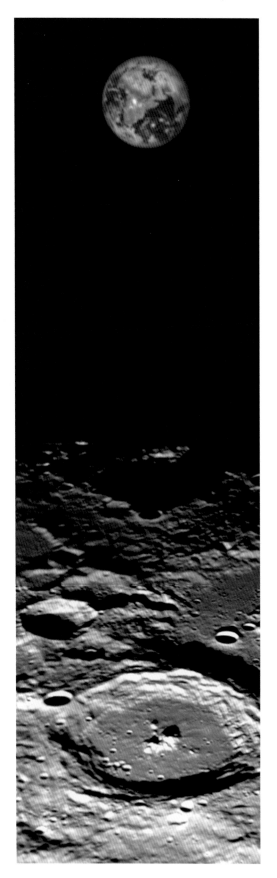

After orbiting the Earth twice, Clementine entered lunar orbit on February 19, 1994. Equipped with several types of cameras, the probe also carried a Laser Image Detection and Ranging Device (LIDAR) for precise cartographic measurements. Capturing more than 25,000 images per day, and thanks to a special image compressor furnished by the French space agency CNES, Clementine data was used to compile an atlas of the Moon from 1,800,000 photos, with resolution down to 100 to 200 meters. For the first time, it was possible to map the Moon's polar regions in great detail, with ground resolution between 15 and 40 meters.

This part of the mission was a total success, particularly with the discovery of ice deposits in craters at the lunar South Pole that are never exposed to sunlight. The probe was subsequently redirected for an encounter with the asteroid Geographos. Unfortunately, due to an on-board computer malfunction, one of the craft's engines used up all available fuel and made it impossible to reach the intended goal, with Clementine stuck in a geocentric orbit. The mission was consequently terminated in June 1994. On the March 5, 1998, NASA officially announced that water ice had been detected by Clementine in certain craters, in sufficient quantities to serve permanent colonies on the Moon. It was also anticipated that enough hydrogen and oxygen could be obtained by electrolysis from this source to be used as rocket fuel. ☾

This celebrated and colorized image of the "full Earth" was taken by Clementine on March 13, 1994, after the probe emerged from under the Moon's North Pole. The crater Plaskett is shown in the foreground.

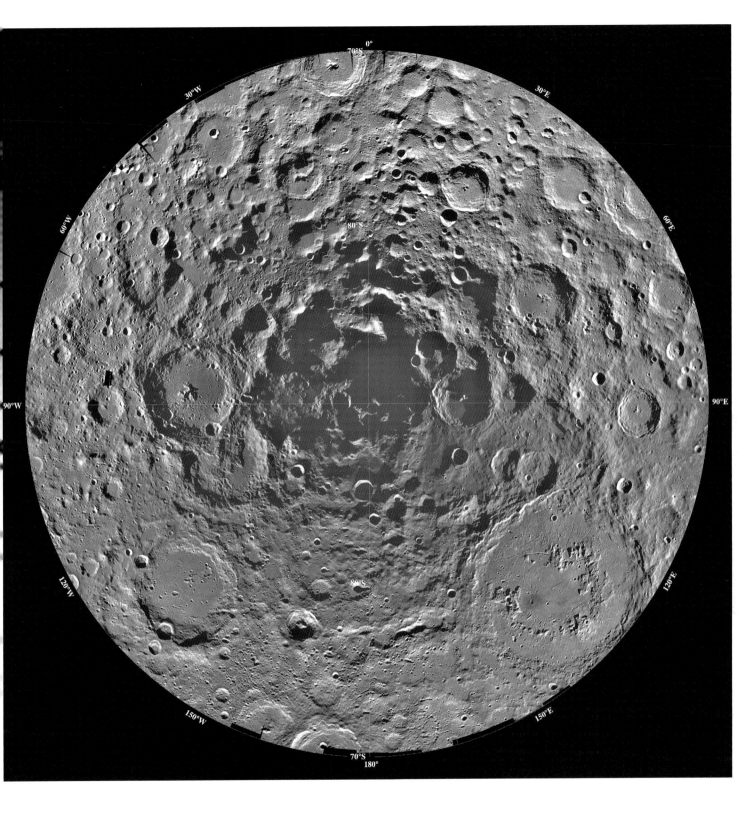

Mosaic of 1,500 images transmitted by Clementine detailing the Moon's South Pole. Several craters at the South Pole appear in permanent shadow. These regions are sufficiently cold to retain ice deposited there by cometary impacts in the past.

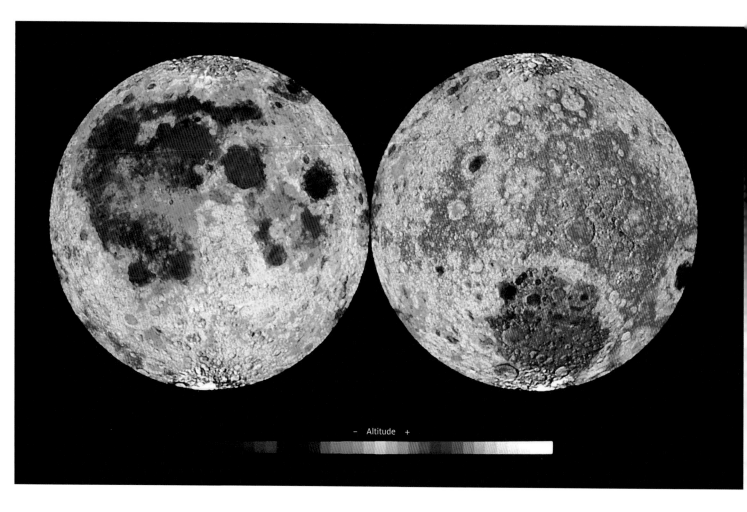

Global topographic map of the Moon from Clementine radar data. The vast depression (2,500 kilometers in diameter) visible at the South Pole of the far side is the Aitken Basin.

– Altitude +

"Oh My Darling, Clementine"

The name of this spacecraft is based on the song "Oh My Darling, Clementine." This popular old folk tune, familiar to most Americans, tells the story of an old miner who was crying because his love had drowned and he was unable to save her because he did not know how to swim. The space probe Clementine was also lost for good once its mission was over, and nothing could have been done to save her.

Lunar Prospector

THE LUNAR PROSPECTOR was one of the Discovery program missions, at a time when NASA's mantra, "Faster, Cheaper, Better," was heavily criticized for the consecutive losses of the Mars Polar Lander and Mars Observer probes. Lunar Prospector, a small, cylindrical craft weighing barely 300 kilograms with fuel, was launched on January 6, 1998, from Cape Canaveral by an Athena 2 rocket. Placed in a polar orbit around the Moon, the mission's main goal was to map the elemental composition of the surface, principally looking for ice deposits and measuring the lunar magnetic and gravitational fields. No camera was on board. The various scientific instruments (several spectrometers, a magnetometer, a reflectometer and a Doppler radio transmitter) were attached to three long radial-support beams, which were deployed in orbit to reduce interference from the spacecraft's central core. Measuring the Moon's gravitational field using the Doppler method helped increase by a factor of 10 the gravitational map obtained by the Clementine mission. This made it possible to detect a 600-kilometer diameter, iron-rich core at the center of the Moon. In addition, the more stable orbit and better fuel economy of Lunar Prospector provided useful information for future lunar missions. This mission confirmed previous findings by Clementine of high concentrations of hydrogen, presumably as part of water molecules, which had accumulated in several polar region craters. To erase any doubts about this, however, mission control decided to crash the probe into the crater Shoemaker, which

A technician examining the Lunar Prospector spacecraft provides a scale as to its size.

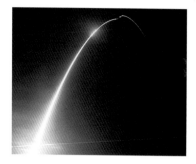

is located near the lunar South Pole. It was hoped that the impact would raise a cloud of water vapor that could be detected by Earth-based spectroscopy. Unfortunately, no spectral signature characteristic of water vapor was detected by Earth-based telescopes. The question of water ice on the Moon consequently remained unresolved. ☾

The launch of Lunar Prospector on January 6, 1998, marked the debut of Lockheed Martin's new Athena II rocket.

BELOW

Artist's rendition of Lunar Prospector, with its instrument beams fully deployed.

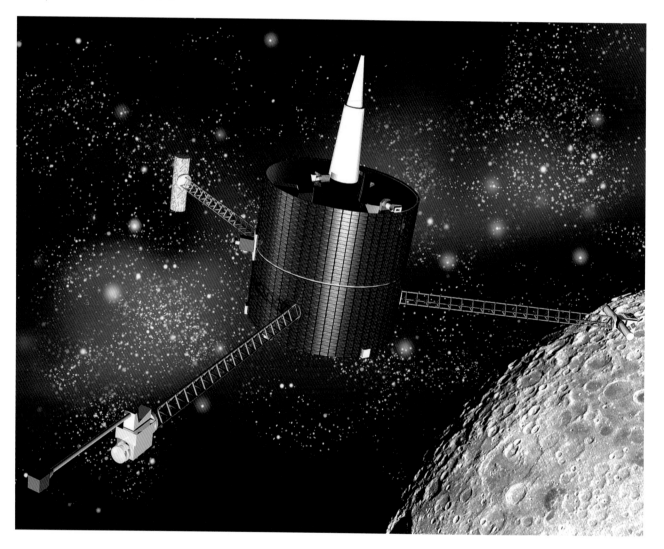

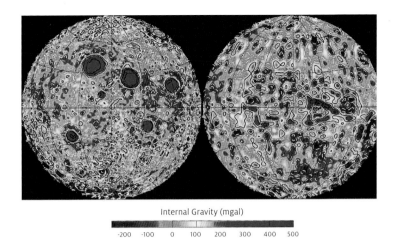

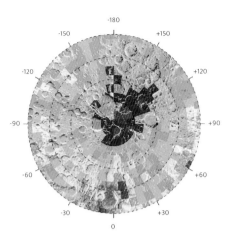

North Pole >70°

Internal Gravity (mgal)

-200 -100 0 100 200 300 400 500

Gravity relief map of the Moon obtained by the orbiting Lunar Prospector, showing vertical variations of different masses.

BELOW

Gravitational and magnetic data obtained by Lunar Prospector helped locate an iron core at the center of the Moon.

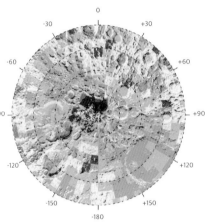

South Pole <-70°

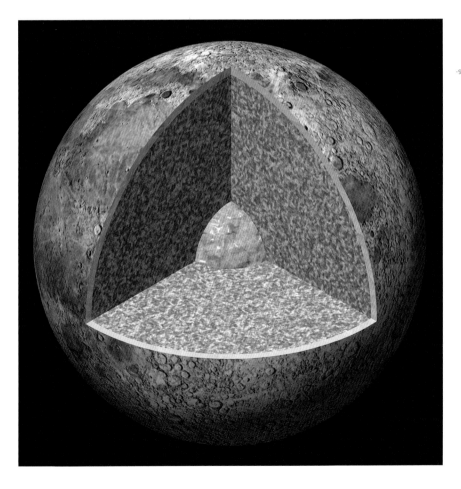

Neutron-emission maps of the Moon obtained by Lunar Prospector revealed significant amounts of water ice (blue areas) in regions of the South Pole that lie in perpetual shade.

SMART-1

The SMART-1 probe being readied in a "clean room."

RIGHT

On September 27, 2003, an Ariane 5 rocket launched the SMART-1 mission into Earth–Moon trajectory from the Kourou launch site of the European Space Agency in French Guyana.

FAR RIGHT

Artist's rendition of the SMART-1 probe in orbit.

SMART-1 was the first European lunar space mission. Its objectives were primarily technological: to test an ion-propulsion system using solar energy and to test miniature electronic components designed to operate in the vacuum of space. The acronym SMART stands for Small Missions for Advanced Research in Technology. The European Space Agency's heavy launch rocket Ariane 5 was chosen to place the SMART-1 probe into a geostationary orbit on September 27, 2003, from the Kourou launch facility in French Guyana.

The first part of the mission, which lasted 16 months, was to test the effectiveness of the ion engine for controlled transfer to a lunar orbit. After following a long spiraling trajectory around the Earth, covering 84 million kilometers, the European spacecraft finally entered lunar orbit on November 15, 2004. Once in orbit, the probe began the second part of its mission, which consisted of

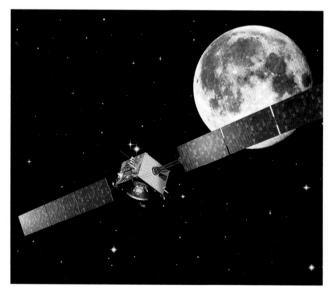

Details of the ion engine that propelled SMART-1. This type of xenon ion engine provided up to 10 times more thrust per kilogram than chemical rockets. Unlike chemical rockets, which are powerful but short-lived, ion engines can operate continuously for months or even years.

BELOW

Trajectory of the circumlunar approach of the SMART-1 mission. More than 16 months were needed before the probe reached its ultimate orbit.

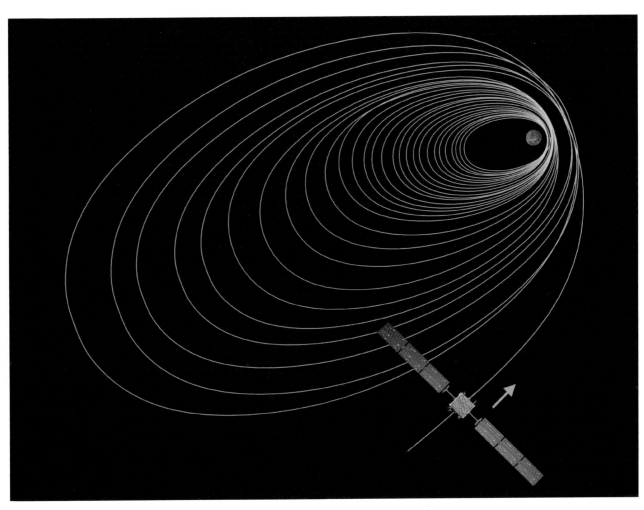

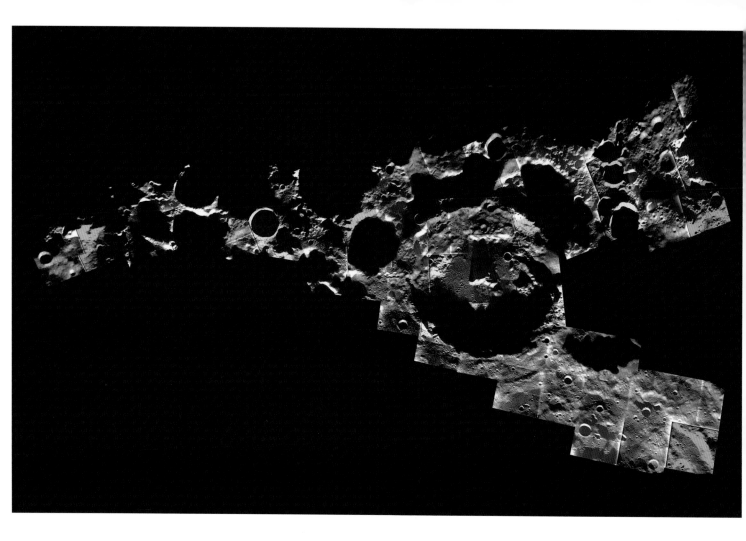

Multiple-image mosaic of a south polar region of the Moon taken by SMART-1. The largest crater shown, Shackleton, located on the edge of the Aitken Basin, has a diameter of 19 kilometers.

studying polar mountain peaks that lie in perpetual sunlight, named "peaks of eternal light" by French astronomer Camille Flammarion, as well as the darkest parts of the lunar poles that might contain ice. SMART-1 was equipped with a miniature digital camera, an X-ray telescope, an infrared spectrometer and a solar activity monitor. The SMART-1 lunar expedition was terminated on September 3, 2006, by deliberately crashing it into one of the dark south polar craters in the Lacus Excellentiae region. Thanks to an advance publicity campaign, the violent impact at a speed of 2 kilometers per second produced a flash that was photographed by several telescopes on Earth. The complete success of SMART-1, both in technological and scientific terms, augured well for future European space missions. ☾

Kaguya (SELENE)

With the launch of two Japanese lunar missions and one by China, 2007 was a year dominated by Asia in the history of astronautics. The space probe SELENE derives its name as an acronym for SELenological and ENgineering Explorer and from Selene, the moon goddess of Greek mythology. It was launched on September 14, 2007, by an H-IIA rocket from the space center at Tanegashima, under the direction of the Japanese Space Agency JAXA. JAXA wanted to emphasize that SELENE (nicknamed Kayuga, the moon princess, by the Japanese public), accompanied by two smaller satellites Okina and Ouna, was the most sophisticated lunar mission since Apollo. After releasing its two satellites, Kaguya entered a near-circular lunar orbit on October 19 at an altitude of 100 kilometers above the surface. Okina served as radio-relay satellite when Kaguya passed "behind" the Moon, while Ouna measured the Moon's gravitational field with high precision through very long base interferometry. Kaguya's scientific equipment included imagers, ground-penetrating radar that could reach down to 50 meters beneath the surface, an altimeter and spectrometers.

At 5 meters tall and weighing 3 tons, Kaguya is an impressive lunar orbiter.

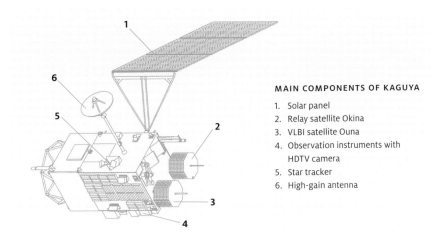

MAIN COMPONENTS OF KAGUYA

1. Solar panel
2. Relay satellite Okina
3. VLBI satellite Ouna
4. Observation instruments with HDTV camera
5. Star tracker
6. High-gain antenna

For public relations purposes, an HDTV camera aboard the spacecraft shot short videos. After the probe's successful scientific mission was achieved in February of 2009, controllers gradually lowered its orbit to 50 kilometers and then to 20 kilometers, and then undertook a directed crash-landing on June 10. Like other space powers, Japan used these spacecraft to position itself for future crewed lunar missions around 2020. ☾

The model H-IIA-2022 rocket was chosen to launch Kaguya on September 14, 2007, from the Tanegashima Space Center in southern Japan.

Close-up view of the two satellites Okina and Ouna, which were placed into lunar orbit to relay signals from Kaguya as it passed behind the moon.

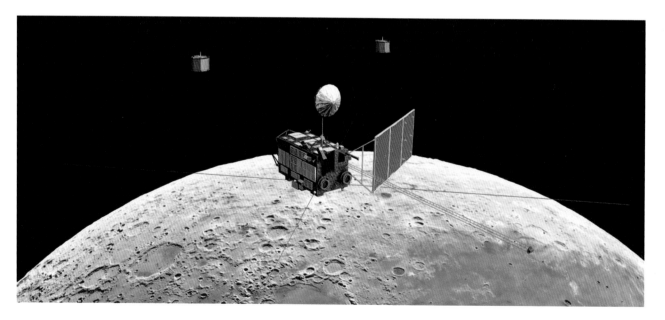

Artist's rendition showing Kaguya and its two accompanying satellites, Okina and Ouna, in interplanetary space between the Earth and the Moon.

The stereoscopic digital camera aboard Kaguya produced this detailed topographic image
of the massive central peak (2 kilometers high) in the center of the crater Tycho.

Kaguya's HDTV camera captured this spectacular earthrise above the Moon.

2007

Chang'e 1 and 2

On October 24, 2007, the Long March CZ-3A rocket, with the Chang'e 1 probe on board, waited for the controllers' green light before firing its engines. The Xichang launching site is mainly used to send satellites into geosynchronous orbit.

RIGHT
Schematic representation of the Chang'e 1 lunar probe.

CHANG'E 1 HAS THE GREAT DISTINCTION of being the first Chinese space probe. The name Chang'e refers to an ancient legend in which a feminine goddess flies toward the Moon and remains there for all eternity. The Chang'e 1 probe, built on the framework of the DFH-3 communication satellite, lifted off on October 24, 2007, carried by a Long March CZ-3A rocket from the Xichang launch site. It orbited the Earth three times, after which it was sent into a translunar trajectory toward its destination. Chang'e 1 entered a stable lunar orbit of 127 minutes at an altitude of 200 kilometers on November 5, 2007. This feat of technology, a first for the Chinese, was celebrated by the radio transmission of 30 folk songs by the spacecraft. The first lunar images became available a few weeks later. In order to fully support their telemetry needs to both track and control spacecraft, the Chinese recently built two enormous parabolic antennas: near Beijing (50 meters diameter) and Kunming (40 meters diameter).

Since China does not have access to a worldwide telemetric network, the European Space Agency (ESA) provided assistance with its antennas, to receive

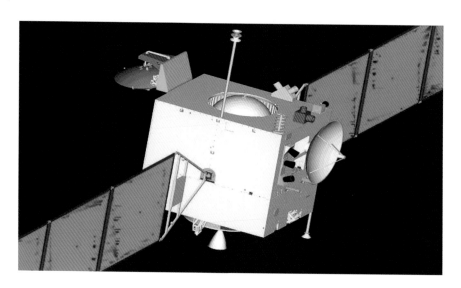

the probe's radio signals, to calculate its coordinates and even to send telecommunication commands. Chang'e 1 carried a stereoscopic CCD camera and a panoply of other measuring instruments, including a laser altimeter and X-ray and gamma-ray spectrometers to determine the amount of useful chemical elements on the lunar surface. The data gathered by Chang'e 1 made it possible to evaluate lunar Helium-3 resources, a rare element on Earth that is in demand for studies on nuclear fusion. Another one of Chang'e 1's goals, the creation of a complete topographical map of the lunar surface, was achieved in November 2008. On March 1, 2009, Chang'e 1 was voluntarily crashed into the Moon. The second Chinese lunar probe, Chang'e 2, identical to its predecessor except for a superior camera, was launched on October 1, 2010, and completed its mission to image the lunar surface in high-resolution (1.5 meters) 3-D from low orbit. Chang'e 1 and 2 established the expertise of the Chinese space agency (CNSA), whose goals are to land rovers and establish an automated base on the Moon before 2020. It seems that the two Asian powers, India and China, are now engaged in a race to the Moon reminiscent of the one between the United States and the USSR in the 1960s. ☾

Rendition of the first 3D photograph of lunar terrain transmitted by Chang'e 1.

BELOW
Complete map of the lunar surface, obtained during the first operational year of the Chang'e 1 mission.

2008

Chandrayaan-1

THE END OF THE 20ᵀᴴ CENTURY augured renewed interest in lunar exploration. This is amply illustrated by Chandrayaan-1 ("Voyage to the Moon" in Hindi), the first Indian lunar mission. It was launched on October 22, 2008, from the Satish Dhawan Space Center, located on Sriharikota Island in the Bay of Bengal. The rocket was an improved version of the polar orbiting satellites (Polar Satellite Launch Vehicle: PSLV), a reliable, four-stage vehicle perfected by the Indian Space Research Organization (ISRO). The Chandrayaan-1 probe was transferred into a polar lunar orbit, on November 12, 2008, at an altitude of 100 kilometers. The payload consisted of a digital camera (TMC for Terrain Mapping Camera) and a laser altimeter, to map the Moon with surface and elevation resolution of 5 to 10 meters. Several of the instruments on board the Chandrayaan-1 orbiter have been provided by foreign collaborators. The spacecraft carried 11 scientific instruments, including two from

Securing the lunar impactor on the Chandrayaan-1 Probe.

OPPOSITE PAGE
Launch of the Indian PSLV rocket.

MAIN COMPONENTS OF CHANDRAYAAN-1

1. Impactor
2. Star tracker
3. Laser altimeter
4. Camera
5. Spectrometric imager
6. Infrared spectrometer
7. Aperture synthesis radar
8. X-ray spectrometer
9. X-ray and high-energy gamma-ray spectrometer
10. Spectral imager
11. Solar panel

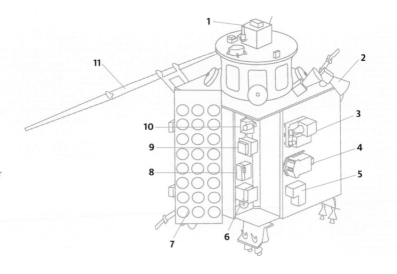

Artist's rendition of the Chandrayaan-1 orbiter mapping mineral resources on the lunar surface.

NASA: a mineralogy spectrometer (Moon Mineralogy Mapper) and synthetic-aperture mini-radar (Mini-SAR) to detect polar ice deposits. ESA's contribution consisted of two high-resolution spectrometers (C1XS and SIR-2) and a subatomic reflection analyzer (SARA). A radioactivity counter constructed by the Bulgarian Science Academy to map levels of radiation in proximity to the Moon was also part of the equipment payload. On November 14, 2008, a small detachable probe, MIP (Moon Impact Probe), was released from Chandrayaan-1 and aimed at a point of impact near the crater Shackleton on the South Pole. The cloud of debris created by the impact's shock was analyzed spectrographically and confirmed the presence of water vapor. The MIP was able to capture a few images of the terrain during the last minutes of descent.

Chandrayaan-1's success demonstrated that India had acquired great confidence in its space technology program since its first satellite in 1975. It has reached a point where it can plan a sophisticated Chandrayaan-2 mission in 2013, in collaboration with Russia, which will involve the deployment of an orbiter and a lunar rover. ☾

STEPS IN THE PROCESS OF POLAR ORBITING OF CHANDRAYAAN-1

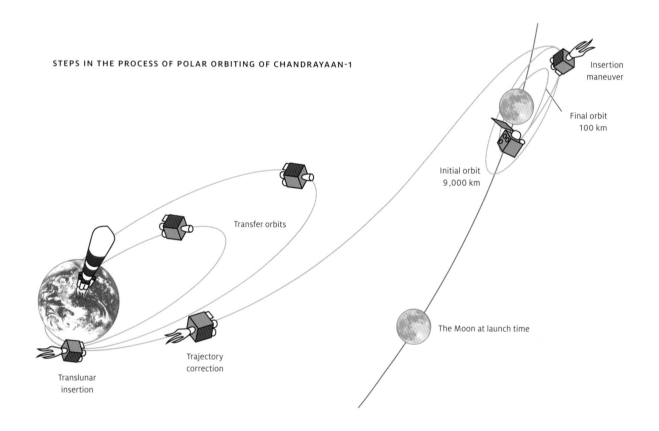

Insertion maneuver

Final orbit 100 km

Initial orbit 9,000 km

Transfer orbits

The Moon at launch time

Trajectory correction

Translunar insertion

Chandryaan-1's high-definition camera capabilities are demonstrated in these images of Earth (left) and the lunar surface (above). The impactor photographed the terrain during the last seconds of its controlled descent.

2009

Lunar Reconnaissance Orbiter and LCROSS

Placement of the LRO and LCROSS probes into the nose cone of the Atlas V Rocket.

OPPOSITE PAGE

Artist's rendition of the orbiting Lunar Reconnaissance probe with the earthrise.

THE FIRST MISSION of NASA's Vision for Space Exploration program, the Lunar Reconnaissance Orbiter-LCROSS, was launched by an Atlas V rocket on June 18, 2009. The probe, placed into a polar orbit, was equipped with several sensors to accurately map the Moon's topography. The main objective was to find the best sites for future crewed landings, estimated to begin in 2020. The high-definition digital images transmitted during this mission resolved objects of 50 centimeters in size and located the rover and LEM modules left behind during the Apollo missions. The orbiter is accompanied by the LCROSS (Lunar CRater Observation and Sensing Satellite) probe. This probe guided the fourth stage of the accompanying Centaur rocket to crash into a polar crater permanently in shadow. By following the Centaur stage four minutes later, the shepherding LCROSS flew through the cloud of debris created by the impact and, before crashing into the Moon, verified the presence of water ice, first indicated by Clementine (see page 67) and Lunar Prospector (see page 71). This mission signaled the return of NASA to the Moon after a 10-year absence, since the Lunar Prospector mission. ☾

MAIN COMPONENTS OF THE LUNAR RECONNAISSANCE ORBITER

1. Cosmic-ray detector
2. Infrared radiometer
3. Miniature radar
4. Deployed solar panel
5. Neutron detector
6. Attitude-control thrusters
7. Stereoscopic cameras
8. Hydrogen detector
9. Laser altimeter
10. Wide-angle camera
11. Star tracker
12. High-gain antenna

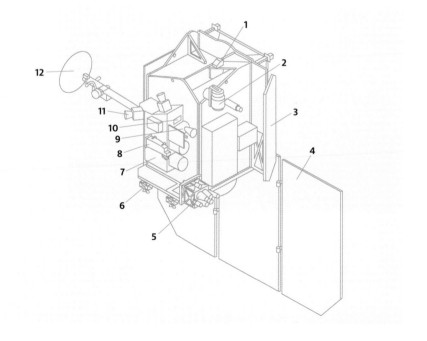

The high-definition images taken by the Lunar Reconnaissance Orbiter allow us to discern the lunar modules (LM) of several Apollo missions and the Surveyor 3 probe (S), which was visited during the Apollo 13 mission.

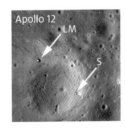

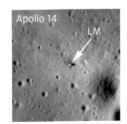

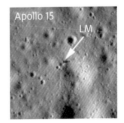

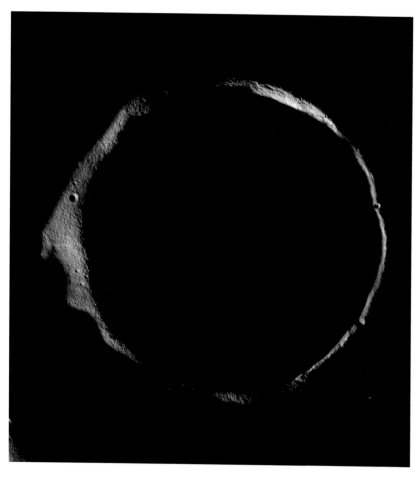

Some examples of the high-resolution mapping capabilities of the lunar surface that have been made possible with the cameras of the Lunar Reconnaissance Orbiter mission. The crater Erlanger (left), located near the North Pole, is an ideal target for radar observations of possible ice trapped in the shadows.

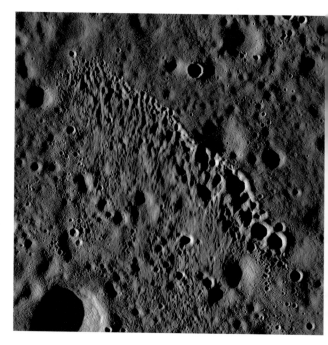

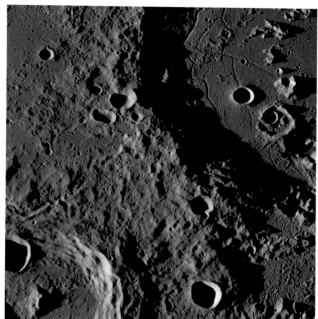

The Centaur rocket crashes into the crater Cabeus on October 9, 2009, and is followed four minutes later by LCROSS probe's crash into the permanently shadowed South Pole region. Spectral analysis of the resultant debris plume clearly indicated the presence of water ice (estimated around 5 percent) in those permanently shadowed areas of our natural satellite.

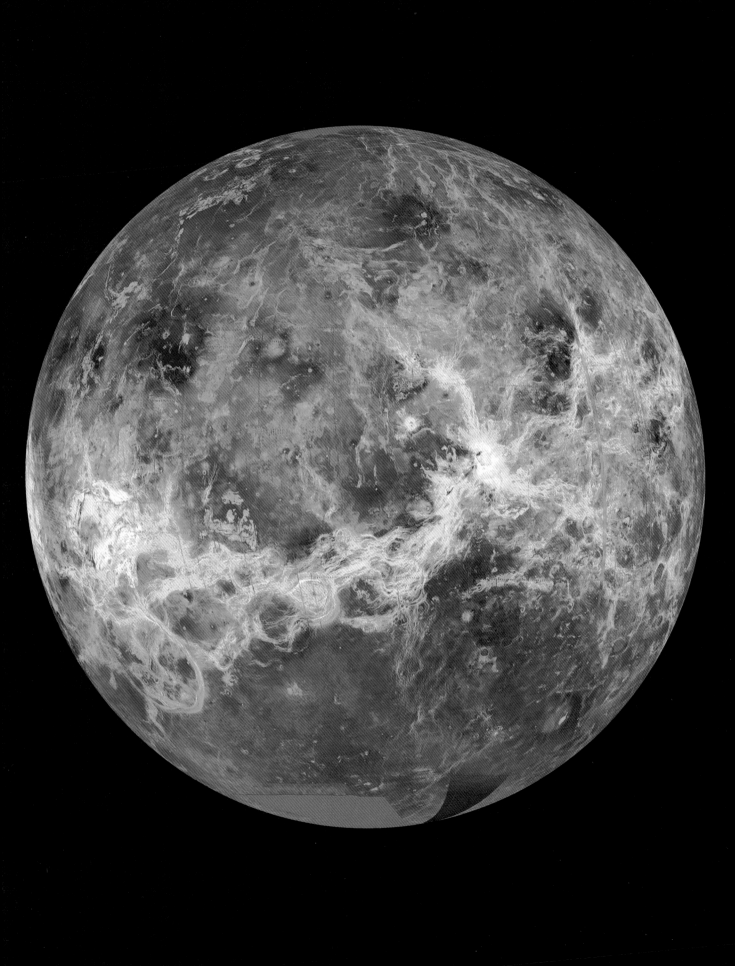

Venus ♀: Our Sister Planet

THE SECOND PLANET of the solar system and the one closest to Earth, the so-called "shepherd's star" is the most brilliant object in the sky after the Sun and the Moon. Once called Hesperus, the morning star, and Phosphorus, the evening star, it now bears the name of the Roman goddess of love and beauty. Its brilliance is due to the fact that the dense clouds that surround it have a high albedo, which is to say that they reflect more sunlight (70 percent) than they absorb (30 percent). By comparison, the Moon reflects only 10 percent of the light it receives from the Sun. Seen from Earth, Venus displays phases similar to the Moon and Mercury. The planet's apparent diameter varies in relation to its phases, since the Earth–Venus distance varies considerably, from 41 to 258 million kilometers.

Venus resembles the Earth in terms of its dimensions, density and probably its internal composition, similarities that warrant its status as our "sister planet." However, in retrospect, the name Venus does not really suit a planet where extreme high pressure exists at ground level and infernally high temperatures render the planet inhabitable. Its dense atmosphere, with ground level pressure 100 times greater than on Earth, is composed mainly of carbon dioxide, nitrogen and sulfuric acid. Because of the high

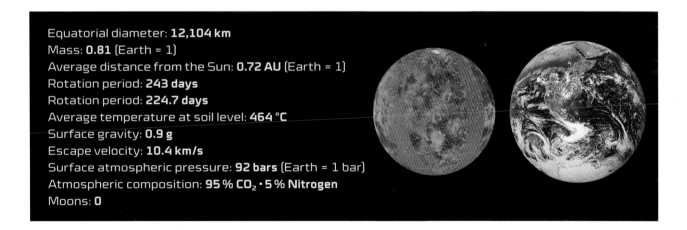

Equatorial diameter: **12,104 km**
Mass: **0.81** (Earth = 1)
Average distance from the Sun: **0.72 AU** (Earth = 1)
Rotation period: **243 days**
Rotation period: **224.7 days**
Average temperature at soil level: **464 °C**
Surface gravity: **0.9 g**
Escape velocity: **10.4 km/s**
Surface atmospheric pressure: **92 bars** (Earth = 1 bar)
Atmospheric composition: **95 % CO$_2$ · 5 % Nitrogen**
Moons: **0**

Variations in the apparent diameter
of Venus during its different phases.
Mercury is similar, as explained in
the diagram below.

concentration of carbon dioxide and clouds in the Venusian atmosphere, the Sun's heat is trapped through the greenhouse effect. Average surface temperatures approaching 500°C were measured by resilient Soviet probes, which makes Venus the hottest planet in the solar system. While permanently hidden by a continuously moving, thick layer of cloud, the planet's surface has been mapped through radar imaging. The presence of rather few impact craters shows that extensive geologic activity has remodeled the terrain for more than 800 million years. The results of the Magellan mission (see page 118) indicate that the surface of Venus is mainly covered by volcanic structures. The International Astronomical Union (IAU) decreed that when mapping the features of Venus, only the names of goddesses or famous women should be used. There is one exception though: Maxwell Montes, the highest summit of Venus, at 11,000 meters, was named after physicist James Maxwell (1831–1879).

On February 12, 1961, the USSR launched Venera 1, the first interplanetary probe in history, which was to come within 100,000 kilometers of Venus. Unfortunately, due to a malfunction of its orientation system, radio contact with the spacecraft was lost a few weeks later. Since this first unsuccessful attempt, some 20 American, Soviet and European space probes have flown by Venus, orbited, peered through its atmosphere and even landed. That does not include the three probes that benefited from a gravitational boost by Venus as they flew by, often at less than 500 kilometers above it: namely Galileo (see page 229) in February 1990; Cassini (see page 238) in April 1998 and June 1999; and Messenger (see page 262) in December 2005 and June 2007. ♀

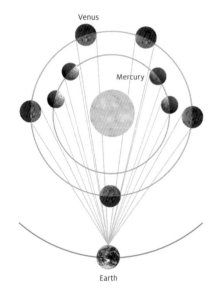

Venus

Mercury

Earth

Mariner 2

I N 1962, THE UNITED STATES definitely lagged behind the Soviet Union in the space race. As yet, however, no well-functioning probe had visited another planet. Two Soviet missions aimed at Mars ended in failure, and the Venera Venusian probes had not reached their goal. The Americans seized the opportunity to demonstrate their know-how. The Mariner 1 and 2 spacecrafts, destined for Venus, were simplified versions of the first Ranger lunar probes (see page 34). Weighing in at 203 kilograms, they consisted of a hexagonal chassis, on which the solar panels, instruments and antennas were mounted. Scientific equipment included two radiometers (microwave and infrared) mounted on a pivoting platform, a micrometeorite detector, solar-wind and plasma detectors, a charged-particle detector and a magnetometer. Since the planet is shrouded in a thick layer of clouds, it was not deemed useful to load on a camera.

The successful liftoff of Mariner 2 atop an Atlas-Agena B rocket on July 22, 1962. Mariner 1was less fortunate, as failure of the booster's guiding system forced its destruction a few minutes after liftoff, for safety reasons.

LEFT

William Pickering (center, left), Jet Propulsion Laboratory's director, and James Webb (center, right), NASA's administrator, present a model of the Mariner probe to President John F. Kennedy in January 1961.

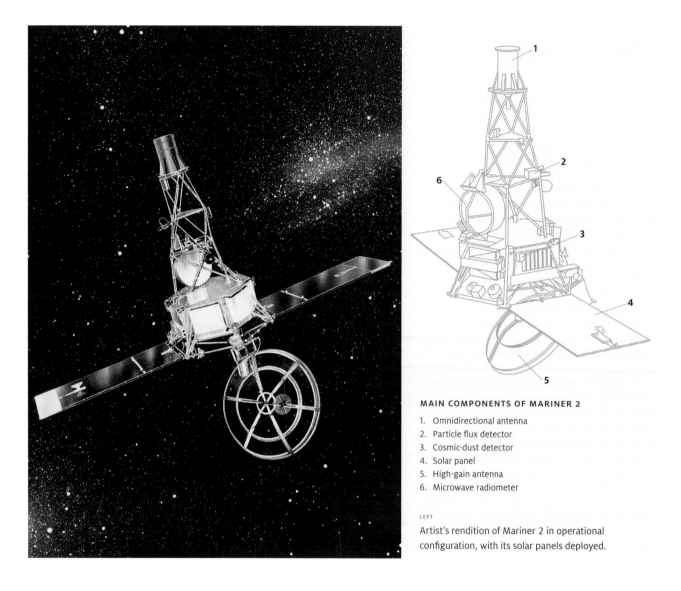

MAIN COMPONENTS OF MARINER 2

1. Omnidirectional antenna
2. Particle flux detector
3. Cosmic-dust detector
4. Solar panel
5. High-gain antenna
6. Microwave radiometer

LEFT

Artist's rendition of Mariner 2 in operational configuration, with its solar panels deployed.

TRAJECTORY TAKEN BY MARINER 2 ON ITS WAY FROM EARTH TO VENUS

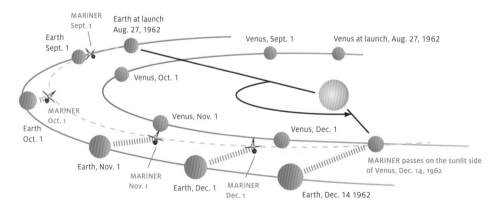

On July 22, 1962, the Atlas-Agena rocket launching Mariner 1 encountered a technical problem and veered from its nominal trajectory. Mission control decided to destroy it in flight less than five minutes after liftoff so that it would not crash into a populated area. Its replacement, Mariner 2, fared better a month later. Its successful launch, on August 27, 1962, propelled it on a three-and-a-half-month voyage toward Venus. During the trip, it confirmed the existence of the solar wind, as indicated by a constant flux of charged ions originating from the Sun. Mariner 2 also observed a brief solar eruption with bursts of high-energy particles and a stream of cosmic rays from sources outside the solar system. On September 8 the probe briefly lost its orientation, but it was quickly re-established, thanks to its gyroscopes. A collision with a small object was the probable cause of this incident, which could have proven fatal to the mission. On November 15, one of the two solar panels abruptly ceased functioning. Fortunately, the probe was close enough to the Sun so that one solar panel was sufficient to power its systems. After a trajectory correction, Mariner 2 flew by Venus on December 14, 1962, approaching it at a distance of 34,830 kilometers. During the flyby, radiometers revealed for the first time that high-altitude clouds on Venus remain cold, while the ground maintained temperatures greater than 200°C, day and night. The probe did not detect a significant magnetic field around Venus. Mariner 2 became the first interplanetary probe in history and the first to fly by another planet. Radio contact with Mariner 2 was lost on January 3, 1963, though it will continue to orbit around the Sun for several thousand years. ♀

Final preparations before launch of Mariner 2, the first interplanetary probe in history.

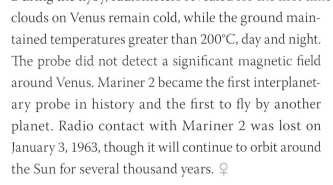

Workers inspect the giant (70 meters in diameter) parabolic antenna of the radio telescope at the Goldstone Space Communications Complex. The complex is operated jointly by NASA, JPL and Caltech. This antenna is part of the Deep Space Network (DSN). It consists of three deep-space communications facilities placed approximately 120 degrees apart around the world: at Goldstone, in California's Mojave Desert; near Madrid, Spain; and near Canberra, Australia. The DSN allows us to receive radio signals (downlink) from interplanetary probes and to communicate with them (uplink), at all times.

Venera 4, 5 and 6

THE SOVIET INTERPLANETARY MISSIONS aimed at Venus were part of the extended Venera (Venus in Russian) program inaugurated on February 12, 1961, with the launching of Venera 1. The latter can be considered as the first interplanetary mission, although its failure has relegated it to oblivion. After it confirmed the presence of solar wind (which had previously been discovered by the Luna probes), radio contact with Venera 1 was lost during the 7-million-kilometer journey from Earth. Because of radio communication problems, no new data were transmitted by Venera 2 and Venera 3, both launched in 1965. Venera 3, which had been launched on November 16, 1965, crashed into Venus on March 1, 1966,

Replica of the Venera 4 lander.

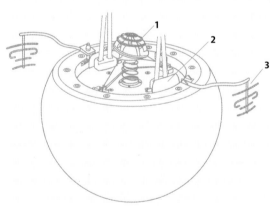

MAIN COMPONENTS OF THE VENERA 4 LANDER

1. Omnidirectional antenna
2. Parachute fastener
3. Radar altimeter antenna (a total of 2)

thereby becoming the first spacecraft in history to reach the surface of another planet. Unfortunately, no data was transmitted. Venera 4 was launched on June 12, 1967, propelled by Molniya, a four-stage rocket derived from the R-7 of the Luna missions. The Soviets once again demonstrated their astronautical expertise, by placing Venera 4 into an Earth parking orbit before firing the fourth stage at the appropriate time and propelling to an encounter with Venus. The mission's goal hinged on a novel technology: the release of a robotic capsule into the Venusian atmosphere, in order to measure atmospheric pressure and analyze its chemical composition, all while on frantic descent by parachute. That goal was attained on October 18, 1967. For the very first time, a space probe had penetrated the atmosphere of another planet in order to study its properties. The descent module of Venera 4 (an oval capsule weighing 384 kilograms) functioned during the 94 minutes of parachute-decelerated fall, transmitting data on the planet's atmosphere and magnetic field. Based on the radio signals and barometric pressure measurements, the probe must have ceased to function at an altitude of 25 kilometers above ground, but not before we learned that carbon dioxide is the main constituent of the Venusian atmosphere, and that it is both toxic and red-hot.

The Soviet spacecraft Venera 4, shown in its orbiting configuration, before the descent module is jettisoned.

In order to withstand the atmospheric pressure during their descent toward the surface of Venus, the lander capsules of the twin missions Venera 5 (launched January 5, 1969) and Venera 6 (launched January 10, 1969) were more solidly built than that of Venera 4. Venera 5 and Venera 6 pierced the Venusian clouds May 16 and 17 respectively, and took a series of measurements of the atmospheric conditions for a bit more than 50 minutes. No image of the surface of Venus, however, had as yet been obtained. ♀

1967 Mariner 5

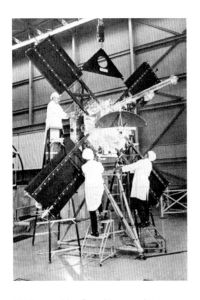

Final assembly of Mariner 5 at the Jet Propulsion Laboratory site in Pasadena, California.

The launch of Mariner 5 on June 14, 1967, from Cape Canaveral on an Atlas-Agena rocket.

M ARINER 5 was a duplicate of the Mariner 4 probe to Mars (see page 132). Since Mariner 4 had accomplished its mission successfully, Mariner 5 was subsequently modified for a mission to orbit Venus. Since Mariner 5 would be going much closer to the Sun than Mariner 4, its thermal insulation was reinforced and smaller solar panels were added. The 245-kilogram probe carried an ultraviolet photometer, a magnetometer, a plasma detector and a radiation detector. One of the main goals of the mission was to conduct radio occultation experiments while the spacecraft passed behind the planet, in order to better measure the thickness and the density of the Venusian atmosphere. Mariner 5 left Cape Canaveral June 14, 1967, propelled by an Atlas-Agena rocket, and flew by Venus on October 19, 1967, at an altitude of 4,000 kilometers and a velocity of 30,000 kilometers per hour. The results of the flyby were first recorded by the spacecraft and then delay-transmitted in batches. The data gathered led to several conclusions: 85 to 99 percent of Venus' atmosphere consists of carbon dioxide; ground-level pressure is estimated at 100 bars; and temperature at 430°C. An experiment to measure the solar wind was coordinated between Mariner 5 and Mariner 4, which remained operational in a heliocentric orbit between Earth and Mars two years after its launch. This way, the existence of a solar wind gradient was established for the very first time. Radio communication with Mariner 5 was permanently lost on November 5, 1968. ♀

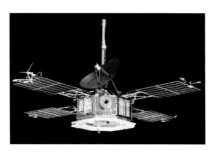

The Mariner 5 probe, initially a duplicate of the Mariner 4 probe to Mars, was adapted for the space environment around Venus.

Venera 7 and 8

O N DECEMBER 15, 1970, Venera 7 became the first human spacecraft to soft land on the surface of another planet and transmit data back to Earth. The Venera 7 mission took off from Baikonur on August 17, 1970, atop a Molniya rocket for a voyage of four months. Similar to that of Venera 4, but larger, the Venera 7 lander was lodged inside an entry capsule in order to survive the extreme conditions on Venus. Its controlled descent lasted 35 minutes, during which the probe measured by Doppler effect that the clouds move at the amazing speed of 400 kilometers per hour. The mission almost ended in disaster when the parachute tore, but thankfully the lander was only a few meters above ground. The probe plummeted in free fall, hitting the surface at a speed of 17 meters per second, bounced once and then rolled onto its side. Since the antenna could not be correctly deployed toward Earth, the quality of the radio signals lessened from then on. The violent landing damaged internal mechanisms and the probe could only measure temperature. Nevertheless, Venera 7 managed to send detectable signals for 23 minutes, despite a temperature of 470°C and 90 bars of pressure.

The Venera 8 mission benefited from the 1972 Venus launch window, lifting off on March 27 and following in the tracks of Venera 7. This time, however, the bus carrying the capsule during the trip was equipped with a refrigeration system, which extended its life after its entry into

Technical inspection of the Venera 7 spacecraft.

Detailed view of the Venera 8 Lander.

BELOW

The Venera 8 lander on the Venusian surface

1. Extended secondary antenna
2. Radar altimeter antenna
3. Atmospheric pressure gauge
4. Photometers
5. Main antenna
6. Secondary antenna prior to extension
7. Parachute

the atmosphere. After reaching the proximity of Venus on July 22, 1972, the bus jettisoned the capsule, which penetrated the atmosphere at a speed of 42,000 kilometers per hour. Atmospheric aerobraking, which is very efficient with Venus, slowed it to 900 kilometers per hour, and a parachute was deployed at an altitude of 60 kilometers. For the first time in the Venera program, the capsule was equipped with a photometer to measure ambient luminosity. The latter recorded a sudden drop at 30 to 35 kilometers above the surface, indicating that a large, thick cloud layer existed at this altitude. Venera 8 landed without problem on the sunlit side of the planet in Vasilisa Regio. Photometric data taken 50 minutes from the surface showed that the light level corresponded to that on Earth during daytime, with 1 kilometer visibility. In addition, the Venera 8 gamma-ray spectrometer indicated rock composition similar to that of granite on Earth. The exploits of the Venera 4 to 8 probes revealed a more precise picture of the Venusian hell, but we still did not have an image of the planet's surface. ♀

Mariner 10

THE LAST MARINER MISSION, Mariner 10, was aimed at an observational flyby of Venus on its way to Mercury. This mission managed to accumulate an impressive series of firsts: first to fly by two separate planets, first to utilize a gravitational sling effect to alter its trajectory, first to return to its objective after a flyby, and first to exploit the pressure of the solar wind as a means of propulsion. Mariner 10 took off from Cape Canaveral in the direction of the Sun on November 3, 1973, propelled by an Atlas-Centaur rocket. Based on the octagonal chassis of Mariner designs, the spacecraft was equipped with several cameras and a numerical registration system, in order to record the images during the planets' flyby and transmit them offline. The cameras were successfully calibrated by aiming at Earth and the Moon during the first week of the Earth–Venus trajectory. An ultra-violet spectrometer, an infrared radiometer and a magnetometer were also part of the utility payload. The main goal of Mariner 10 was to carry out the first close-up study of Mercury. The Italian scientist Giuseppe "Bepi" Colombo (1920–1984) proposed an ingenious maneuver that exploited Venusian gravity to direct the probe's trajectory toward Mercury. This maneuver also saved on the heavier fuel load that would have been necessary for a direct Earth–Mercury trajectory. The Mariner 10 trip to Venus had its own set of problems. Pieces of bright paint chips that became detached from the probe

Assembling the Mariner 10 probe before transport to the launch site.

confused the star trackers aimed at Canopus, resulting in loss of orientation several times. In addition, the on-board computer experienced several breakdowns, requiring ground control to reprogram it and reconfigure the flight path. When the spacecraft closed to within 5,746 kilometers above Venus on February 5, 1974, it began to observe the planet in visible and ultraviolet light. To everyone's surprise, images of Venus, shot through an ultraviolet filter, showed that various cloud bands rotate around the planet. After completing the Venusian leg of its mission, Mariner 10 continued its voyage to Mercury (see page 259). ♀

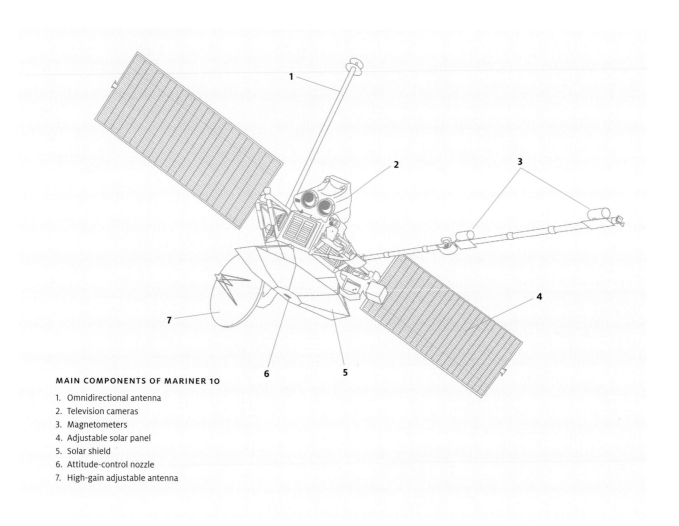

MAIN COMPONENTS OF MARINER 10

1. Omnidirectional antenna
2. Television cameras
3. Magnetometers
4. Adjustable solar panel
5. Solar shield
6. Attitude-control nozzle
7. High-gain adjustable antenna

Two views of Venus: in visible light (left) and in ultraviolet light (right), as transmitted by Mariner 10.

LEFT

Trajectory of Mariner 10 to Venus and Mercury, as calculated by the Italian astronomer Giuseppe "Bepi" Colombo to provide for several Mercury flybys.

Earth orbit

Launch
Nov. 3, 1973

EARTH

Mercury orbit

VENUS   Venus flyby
Feb. 5, 1974

Mariner 10 orbit

Venus orbit

SUN

MERCURY
Mercury flyby
Mar. 29, 1974
Sept. 21, 1974
Mar. 16, 1975

1975   Venera 9 to 14

Replica of a typical Venera 9 to 14 lander.

RIGHT
Overall view of the Venera 9 probe in orbital configuration, before release of the lander in its protective sphere.

OPPOSITE PAGE, TWO TOP PHOTOS
The first photographs of the surface of another planet taken by a lander. Despite severe operating conditions, the Soviet probes Venera 9 and Venera 10 accomplished this historic feat on Venus in October 1975.

OPPOSITE PAGE
Mosaic of landing site photographed by the panoramic cameras on Venera 13 (in color) and Venera 14 (the two lower photos). Some Venusian rocks appear similar to terrestrial basalts, judging by their shape and chemical composition. No sign of erosion is evident in these images.

AFTER GAINING SOLID EXPERTISE in missions to Venus, the Soviets decided to send two spacecraft with landing capability to the planet, so as to return the very first images of the surface. The twin missions, Venera 9 and 10, sent during the launch window of 1975, consisted of an orbiter (a first on Venus) and a lander. In addition to observing atmospheric phenomena, each orbiter acted as a radio-relay system to transmit data from the landers to Earth. The spacecraft's considerable mass, nearing five tons, required powerful launchers — in this case, the Proton rockets. Venera 10, weighing 5,033 kilograms, was launched May 14, 1975, and Venera 10, with a mass of 4,936 kilograms, followed it on June 8. Because of differences in trajectories, Venera 9 reached Venus first, on October 20. The descent maneuver toward the planet began on October 22.

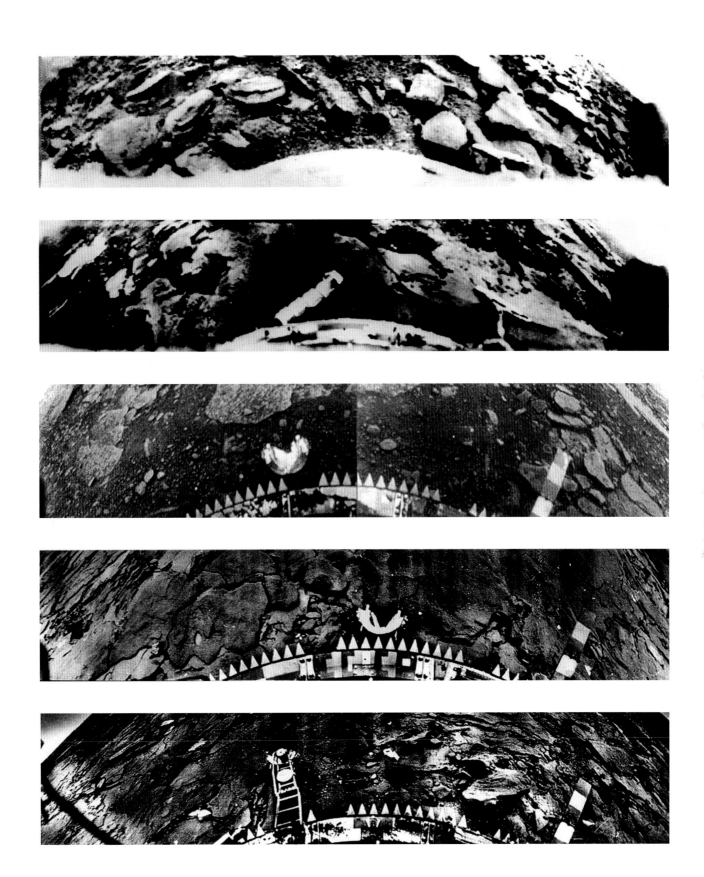

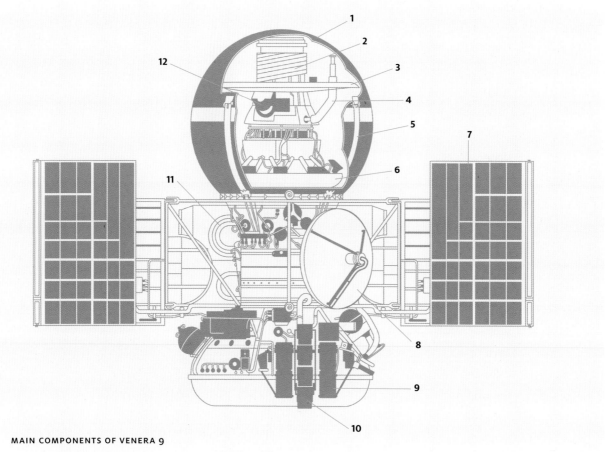

MAIN COMPONENTS OF VENERA 9

1. Lander parachute compartment
2. Coiled antenna for communication between lander and orbiter
3. Aerodynamic brake
4. Heat dissipation system
5. Parachute cord
6. Shock-absorbing ring
7. Solar panel
8. High-gain antenna
9. Star tracker
10. Sun tracker
11. Radiator
12. Photometric cameras (two)

The 1,560-kilogram lander was slowed down sequentially during its descent through aerobraking in the upper layers of the Venusian atmosphere, made possible by a thermal protection aeroshell, and by deployment of three parachutes assisted by a ring-shaped drag brake. A compressible metal cushion served as shock absorber during final impact. Venera 9 landed on a 20-degree incline in the Aikhulu Chasma valley and transmitted images of its surroundings. This made it the first space probe in history to transmit images from the surface of another planet. The Venera 9 landscape showed rocks 30 to 40 centimeters in size, without any traces of erosion or dust.

Venera 10 landed three days later, 2,200 kilometers from Venera 9 in the southeast section of Beta Regio. At this location, surface rocks appeared flat and basaltic. During the two missions, the cover of one of the two cameras remained locked in closed position, which made it impossible to complete a 360-degree panoramic view of the site. Thanks to a system of lines with circulating liquid refrigerant, the landers were able to remain active under the extreme surface conditions for extended periods of time. Venera 9 lasted for 53 minutes, and Venera 10 survived for 65 minutes.

The following four Venera missions, including Venera 11 (launched September 9, 1978) through Venera 14 (launched November 4, 1981), were similar in structure, but were scheduled to fly by Venus, after the jettisoning of their landers, rather than staying in orbit around the planet. Venera 11 and 12 were able to observe lightning and thunder during their descent. Although Soviet engineers tried hard to fix the problem of the camera covers, unfortunately, their new design was worse than the original: the Venera 11 and 12 landers could not transmit any images. Finally a definitive solution was found, and Venera 13 and 14 returned the first color images of their landing site. Nevertheless, nothing working perfectly, one camera cover on Venera 14 managed to block the apparatus to measure soil compression!

Artistic interpretation of the surface of Venus with lightning, as was often detected by the instruments on several Venera probes.

These missions confirmed that a layer of 30 to 40 kilometers of clouds revolves around our sister planet, at an altitude of 30 to 35 kilometers. Since there is no dust, ground level visibility on Venus is excellent. The Venusian rocks seem to have a composition similar to terrestrial volcanic rocks, which are composed of granite and basalt. The chemical analyzers also detected hydrochloric acid, hydrofluoric acid, bromide and iodine in the less than hospitable Venusian atmosphere.

The Soviet space program met with much more success with its Venus landers than with those sent to Mars (no success yet). This could be explained by the fact that the descent process in the very dense Venusian atmosphere requires less active control of the altitude than in the thin Martian atmosphere. ♀

Pioneer Venus Orbiter

Final inspection of the Pioneer Venus Orbiter.

Artistic interpretation of the probe orbiting Venus.

AFTER THE SHORT FLYBY of Mariner 10 in 1973, the Americans more or less neglected Venus. Although Soviet probes had landed there many times, overall mapping of the surface had still not been done. In order to map Venus, it was necessary to penetrate the opaque layer of clouds that shrouded the surface, and that could only be done with radar. Radar, by sending and receiving radio waves to the planet's surface, can penetrate clouds or fog and detect solid objects, determining distance or speed. The first Venus mapping mission using radar was undertaken by the Pioneer Venus Orbiter.

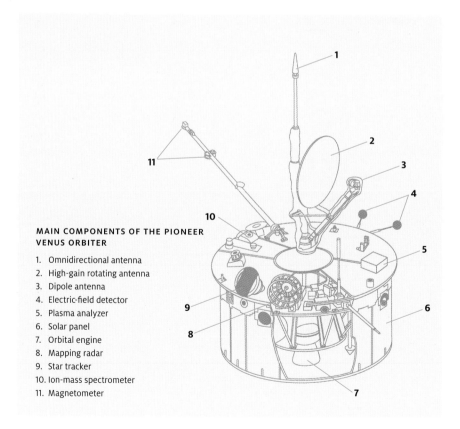

MAIN COMPONENTS OF THE PIONEER VENUS ORBITER

1. Omnidirectional antenna
2. High-gain rotating antenna
3. Dipole antenna
4. Electric-field detector
5. Plasma analyzer
6. Solar panel
7. Orbital engine
8. Mapping radar
9. Star tracker
10. Ion-mass spectrometer
11. Magnetometer

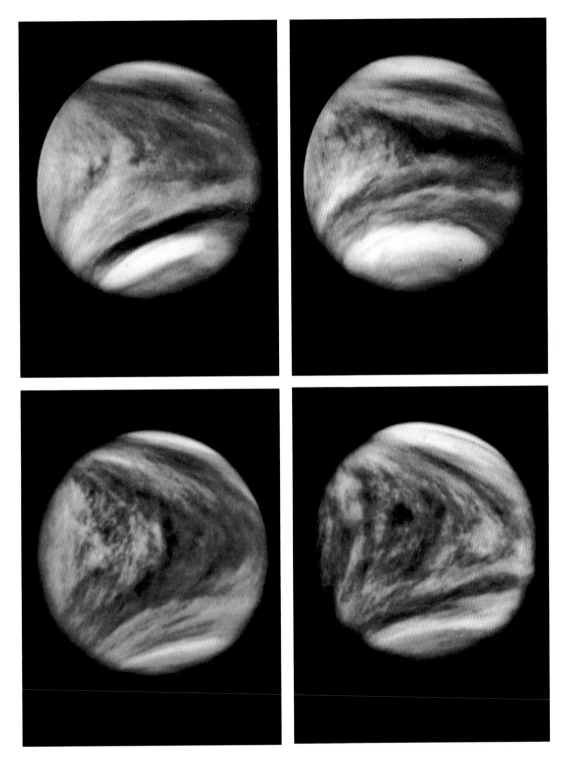

Images of the upper layers of the Venusian atmosphere, taken through an ultraviolet filter, show the rapid rotation ("super-rotation") of clouds around the planet during a four-day period. The clouds move at the astonishing speed of 100 meters per second!

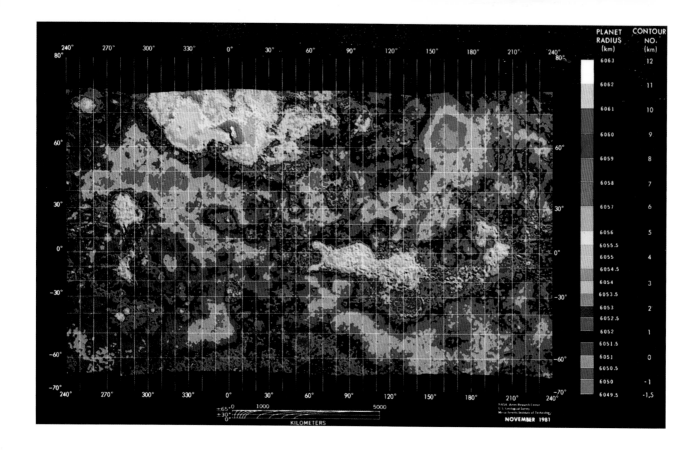

PLANET RADIUS (km)	CONTOUR NO. (km)
6063	12
6062	11
6061	10
6060	9
6059	8
6058	7
6057	6
6056	5
6055.5	
6055	4
6054.5	
6054	3
6053.5	
6053	2
6052.5	
6052	1
6051.5	
6051	0
6050.5	
6050	-1
6049.5	-1.5

NASA Ames Research Center
U.S. Geological Survey
Mass. Inverts Institute of Technology
NOVEMBER 1981

First global map of the topography of the planet Venus. Vast plains cover 60 percent of its surface. Aphrodite Terra to the north and Ishtar Terra at the equator form two continents. The map has a resolution of 75 kilometers (the size of the smallest element).

Pioneer Venus Orbiter was launched from Cape Canaveral on May 20, 1978, by an Atlas-Centaur rocket. It was inserted into an elliptical orbit around Venus on December 4, with perihelion (minimum altitude) between 142 and 253 kilometers and aphelion (maximum altitude) of 66,900 kilometers. This was done to let it circle Venus as long as possible to cover the entire surface, a goal realized in July 1980. In addition to the radar, which also served as an altimeter, the probe was equipped with a battery of other measuring instruments. Seventeen experiments were conducted in total, many to specifically characterize atmospheric properties. The distribution of clouds was evaluated with a photo-polarimeter, and their infrared emission measured by a radiometric mass spectrometer. The first phase of the Mariner 10 mission concluded when the orbiter was redirected into a perihelic orbit, 2,290 kilometers above the surface (to conserve fuel), and the radar mapper was inactivated for future use. It was re-activated in 1991 to work in tandem with the Magellan mission (see page 118). In May of 1992, the orbiter was lowered to a perihelion of 150 to 250 kilometers and kept there until the fuel ran out in August. ♀

Pioneer Venus Multiprobe

I N ORDER TO OPTIMIZE RETURNS in the 1978 launch window to Venus, the Pioneer Venus Orbiter probe was accompanied by the Pioneer Venus Multiprobe, which was launched a few months later on the August 8. The Pioneer Venus Multiprobe was to enter the atmosphere with several smaller capsules and spread over different zones on Venus. The objective was to study the structure and composition of the atmosphere down to the surface, the nature and composition of the clouds, the radiation field and energy exchange in the lower atmosphere, and local information on the atmospheric circulation pattern.

Structurally, the Multiprobe mission consisted of five elements: a carrier bus spacecraft and four atmospheric probes (one large one and three small ones — all identical). The probes carried a series of measuring instruments (as did the bus, which burned up due to atmospheric friction) and were pressurized to resist high Venusian pressures. These probes were not intended to land and therefore did not carry either cameras or soil-analysis instruments. Their survival on the surface of the planet was optional and viewed as a bonus. Only the large probe was equipped with a parachute and

Inspection of the Pioneer Venus Multiprobe spacecraft.

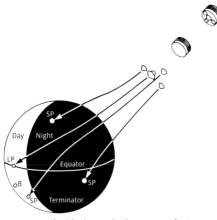

Sequence of critical steps in the entry trajectory into the Venusian atmosphere of the three small probes (SP), the large probe (LP) and the bus (B) of the Multiprobe mission.

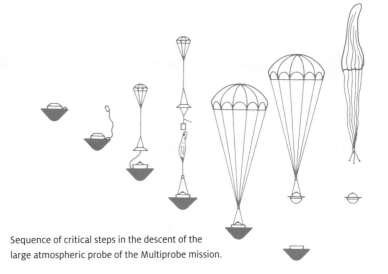

Sequence of critical steps in the descent of the large atmospheric probe of the Multiprobe mission.

a detachable thermal shield. On November 16, 1978, the bus jettisoned the large probe, and on November 20, the three small probes. The four projectiles penetrated Venus' atmosphere on December 9, which is five days after the Pioneer Venus orbiter was set into orbit. Depending on the zone of the planet that they were visiting, the three small probes were named North, Day and Night. The Day probe survived the landing impact and transmitted signals for more than one hour from the surface of Venus, while the others crashed. All in all, the descent had lasted about 50 minutes. Thanks to these two Pioneer Venus missions, our knowledge of our sister planet jumped qualitatively in 1978. ♀

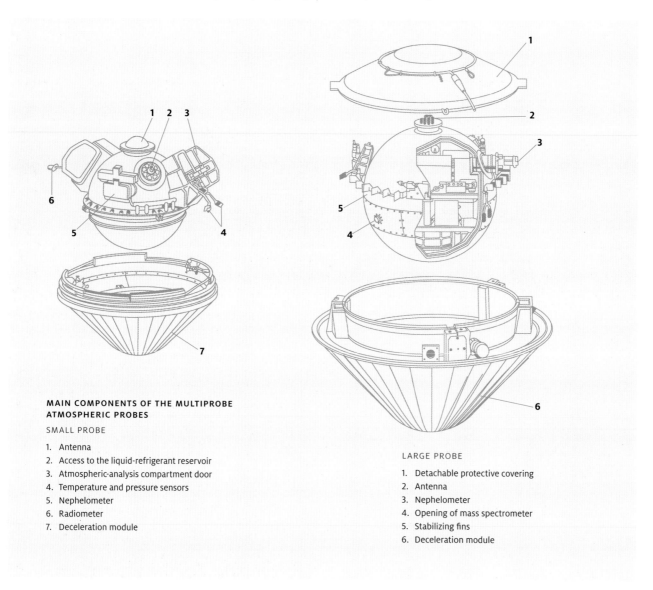

MAIN COMPONENTS OF THE MULTIPROBE ATMOSPHERIC PROBES

SMALL PROBE

1. Antenna
2. Access to the liquid-refrigerant reservoir
3. Atmospheric-analysis compartment door
4. Temperature and pressure sensors
5. Nephelometer
6. Radiometer
7. Deceleration module

LARGE PROBE

1. Detachable protective covering
2. Antenna
3. Nephelometer
4. Opening of mass spectrometer
5. Stabilizing fins
6. Deceleration module

Venera 15 and 16

T HE OBJECTIVE OF THE TWIN Venera 15 and 16 missions was to radar map Venus from an orbital position, providing better resolution than the Pioneer Venus Orbiter. Consequently, the atmospheric re-entry capsule of the preceding Venera missions was replaced on Venera 15 and 16 by a synthetic-aperture radar antenna. This type of sweeping radar is a technique that greatly increases the precision of topographical readings. A parabolic antenna, pointing in the same direction as the radar antenna, served as a radio altimeter

Venera 15 being readied for launch with its closed radar (left), and in its mapping configuration with its reflecting radar in an open position (right).

The combination of altimetry data and radar images from the Venera 15 and 16 missions improved the quality of topographic information, shown here in the region of Ishtar Terra.

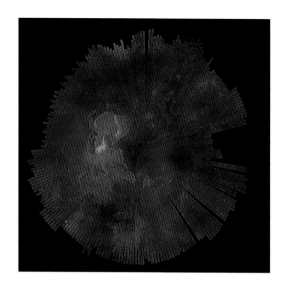

MAIN COMPONENTS OF VENERA 15

1. Radar antenna
2. Radar altimeter antenna
3. Radar module
4. High-gain communication antenna
5. Solar panel
6. Radiator
7. Star tracker
8. Nitrogen tank for attitude control
9. Refrigeration circuit

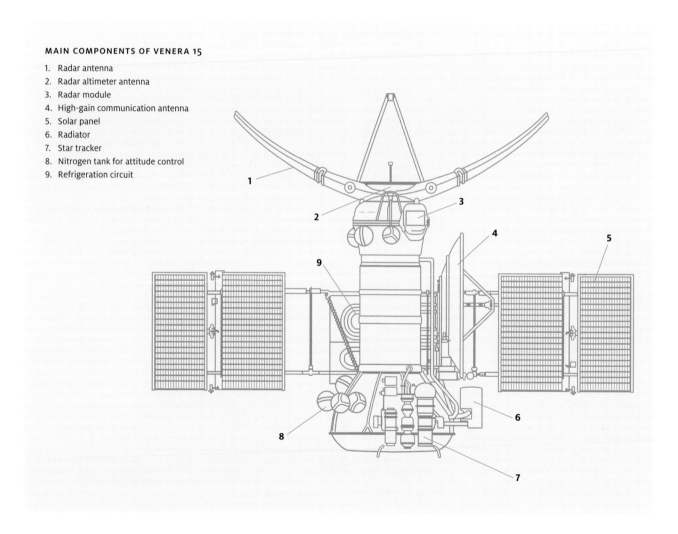

to measure elevations. Venera 15 lifted off from Baikonur on June 2, 1983, on a Proton rocket, and Venera 16 on June 7. They entered a 1,000 by 65,000 kilometers elliptical polar orbit, on October 10 and 14 respectively, so as to cover areas between the North Pole and 30°N latitude, or 25 percent of the Venusian surface. By occupying two close orbits, the same area could be covered twice if necessary. Each probe carried powerful computers to reduce the radar data before transmitting the images. Each radar image, with resolution of 1 kilometer, covered a narrow band 120 kilometers wide by 7,500 kilometers long and took 16 minutes to complete. In June of 1984, Venus entered superior conjunction: that is to say, it was behind the Sun in relation to Earth, meaning that all communication was blocked. Venera 16's orbit was therefore modified to pass over the areas missed during conjunction. The mission continued to July 1984. In the areas covered by Venera 15 and 16, the planet was shown to be more geologically active than Mars, but without the tectonic plates found on Earth. ♀

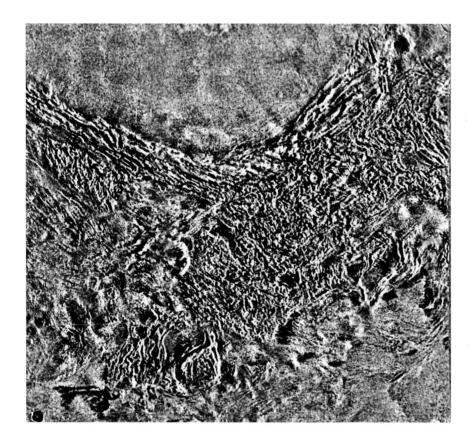

Despite the poor resolution of the transmitted radar images, intense geologic activity on the surface of Venus is evident.

Vega 1 and 2

THE SOVIET VEGA PROGRAM'S MISSION was to study the planet Venus by taking advantage of the passage of the comet Halley at its perihelion in 1986. The name of this ambitious expedition came from the contraction of the words VEnera and GAllei (Halley in Russian). Two identical space-crafts of the Venera generation, the massive Vega 1 and 2 (4 tons), were to fly by Venus, jettison a lander and then direct themselves toward Halley's Comet. Vega 1 was launched by a Proton rocket on December 15, 1984, and Vega 2 on the 21st of that month.

The two probes passed in the neighborhood of Venus in June 1985. Vega 1 freed its 750-kilogram descent module, which consisted of an identical lander to those of the preceding Venera missions, and a balloon with a payload basket full of instruments to study the atmosphere. This package was designed in collaboration with a French team from the CNES (Centre National d'Études Spatiales). The assembly penetrated the atmosphere at 11 kilometers per second, but due to a strong blast of wind that released the impact sensors prematurely, the lander deployed its instruments during the descent. It crashed without transmitting any data. The module containing the balloon and its basket was separated from the lander by opening a parachute at an altitude of 61 kilometers, 40 seconds after entering the atmosphere. The opening of a second parachute released a balloon that stabilized at an altitude of about 54 kilometers, a turbulent area where the pressure is 535 mbar and the ambient temperature 40°C. The horizontal displacement speed of the balloon probe was estimated at 69 meters per second, and on June 12 it passed the terminator zone, or the night-day transition zone. This atmospheric probe drifted for a distance of 11,600 kilometers and then stopped communicating on June 13, 1984.

Moving the Vega 2 lander.

For its part, Vega 2 released its descent module on June 15, 1985, which set down without problems in the northern region of Aphrodite Terra. Analysis of surface samples indicates the presence of rocks of the anorthosite-troctolite variety. This type of rock formation is rarely found on Earth, but is common on lunar highlands. Since it is indicative of very ancient material, it suggests the Vega 2 landing site is the oldest so far visited on Venus. After operating for 56 minutes, the lander succumbed to the infernal Venusian conditions. The data from the Vega 2 balloon probe, including the information about atmospheric conditions, are similar to those obtained by Vega 1. The Venusian phase of the Vega program concluded on June 17, 1985.

The next phase, to investigate Halley's Comet, began after the trajectories of both Vega probes were deflected through the Venusian gravitational field. ♀

The concept of a balloon probe had several advantages when it came to the exploration of Venus, especially its high speed of displacement in the atmosphere. A Kevlar balloon prototype is shown undergoing testing for future missions, at JPL in Pasadena.

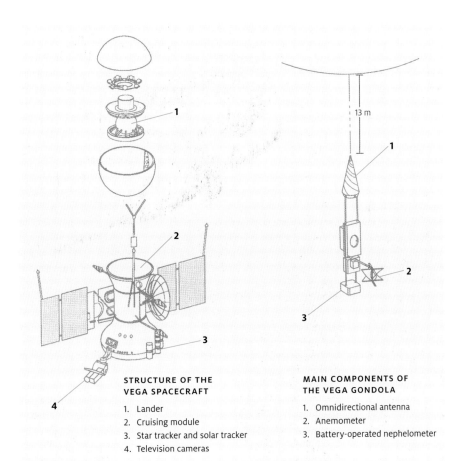

13 m

STRUCTURE OF THE VEGA SPACECRAFT

1. Lander
2. Cruising module
3. Star tracker and solar tracker
4. Television cameras

MAIN COMPONENTS OF THE VEGA GONDOLA

1. Omnidirectional antenna
2. Anemometer
3. Battery-operated nephelometer

1989   **Magellan**

BELOW, LEFT

Preparation of the Magellan probe at Martin Marietta Astronautics in Denver, Colorado.

BELOW, RIGHT

Liftoff of the shuttle Atlantis on May 4, 1989, for the STS-30 mission, whose main goal was to place Magellan in a low Earth orbit at 296 kilometers, before its launch toward Venus by a two-stage rocket.

THE 1989 MAGELLAN MISSION, named in honor of the famous Portuguese explorer, was the first American interplanetary mission since Voyager in 1977 (see page 219). Destined to completely map the surface of the planet Venus in detail, the Magellan orbiter was not launched by a conventional rocket. It was the first of three space probes that were transported into Earth orbit by a space shuttle before being sent to their respective targets (see Galileo page 229 and Ulysses page 278). The shuttle Atlantis lifted off on May 4, 1989, from Cape Canaveral with Magellan as its payload. The astronauts

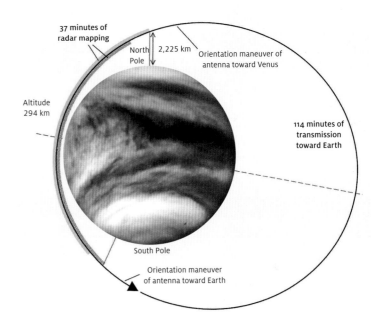

37 minutes of radar mapping

2,225 km

North Pole

Orientation maneuver of antenna toward Venus

Altitude 294 km

114 minutes of transmission toward Earth

South Pole

Orientation maneuver of antenna toward Earth

Deployment of Magellan from the shuttle Atlantis in orbit.

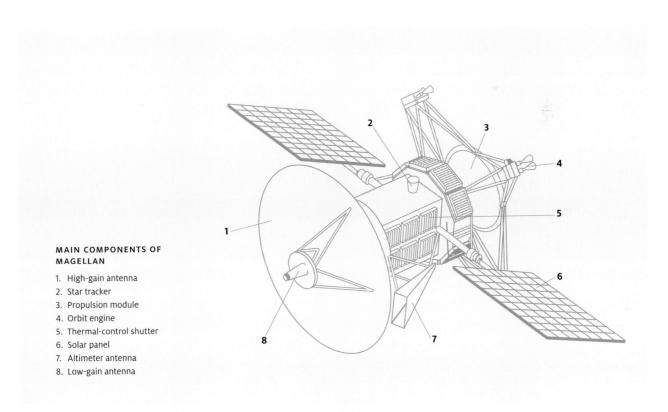

MAIN COMPONENTS OF MAGELLAN

1. High-gain antenna
2. Star tracker
3. Propulsion module
4. Orbit engine
5. Thermal-control shutter
6. Solar panel
7. Altimeter antenna
8. Low-gain antenna

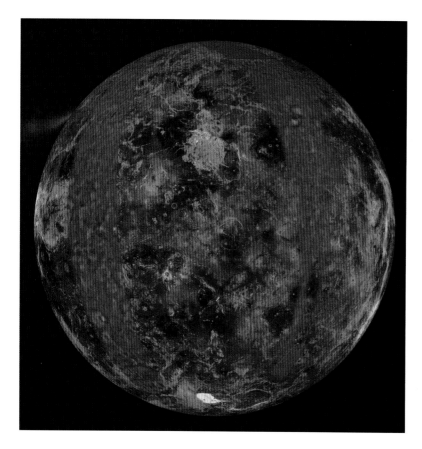

Global view of the surface of Venus, based on altimetry and radar data from Magellan. The view is centered on the equator. The color is based on ground-level observations from the Venera 13 and 14 missions.

RIGHT

This volcanic structure, nicknamed a "tick," refers to a caldera with radial outpourings of lava. This particular volcanic caldera, 30 kilometers in diameter, is located in the northeast zone of Alpha Regio.

FAR RIGHT

These pancake-like volcanic domes in Alpha Regio were probably formed by the extrusion of high-viscosity lava, which would account for their sharp edges.

of the STS-30 mission carefully released the probe from the payload bay and left it in a low orbit, 296 kilometers above Earth. Following this, the probe's IUS (Inertial Upper Stage) solid-fuel rocket was ignited to launch it in a long trajectory, requiring two passes around the Sun before it approached Venus. A stronger push would have shortened Magellan's voyage, but for safety reasons, the size of the IUS motor was reduced. Mission control was nervous about transporting large quantities of fuel in the shuttle payload. On August 10, 1990, Magellan settled in an orbit around Venus at an altitude varying between 250 kilometers and 8,070 kilometers. To limit mission costs, the probe was made of several recycled items from previous missions, including Voyager (high-gain radio antenna and retrorockets), Galileo (on-board computer and electrical power subsystem) and even Mariner 9 (moderate-gain radio antenna). Magellan's high-gain antenna served for both radar imaging and communications with Earth. As with the Venera 15 and 16 missions (see page 113), the radar technique known as synthetic aperture was used for cartography and gravimetry, but on the Magellan mission it was used more systematically and with greater precision, to cover almost the whole planet.

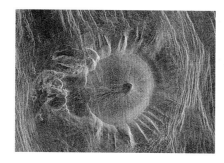

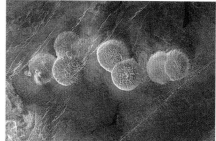

The mission was divided into several phases of 243-day imaging cycles (which correspond to Venus' rotation period) and lasted from 1990 to 1994. The effectiveness of aerobraking was first appreciated when Magellan had to be lowered to an altitude of 180 kilometers for gravimetric mapping.

On October 11, 1994, after completing some 15,000 orbits around Venus, mission control lowered the probe into the planet's upper atmosphere, where it quickly disintegrated. During its four years of operation, the Magellan mission was a complete technical and scientific success. The spectacular radar images returned by Magellan are of such high quality that they are on a par with those made with classic photographic methods. The mission mapped 98 percent of the surface of Venus, with a resolution of approximately 100 meters. With 22 percent of the images in stereoscopic mode, the data produced an unequaled resource for understanding the geology of this planet. Unlike the Earth, Venus does not have tectonic plates, and its internal heat can only escape through destructive volcanic activity that constantly remodels the surface of this planet. ♀

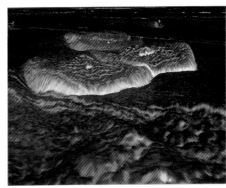

A 3-D reconstruction of several of the pancake-shaped domes characteristic of Venus.

BELOW

This relief map of Venus, the most complete to date, is based on data from the Magellan mission. In the absence of oceans to define a base level, the elevations shown correspond to altitudes measured relative to the minimum radius of the planet (6,048 kilometers).

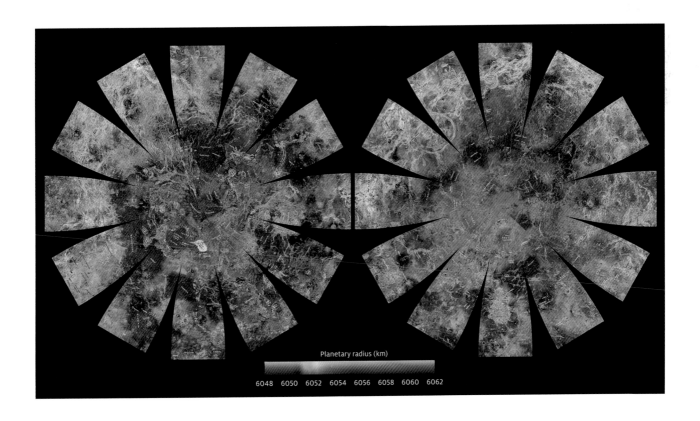

Planetary radius (km)

6048 6050 6052 6054 6056 6058 6060 6062

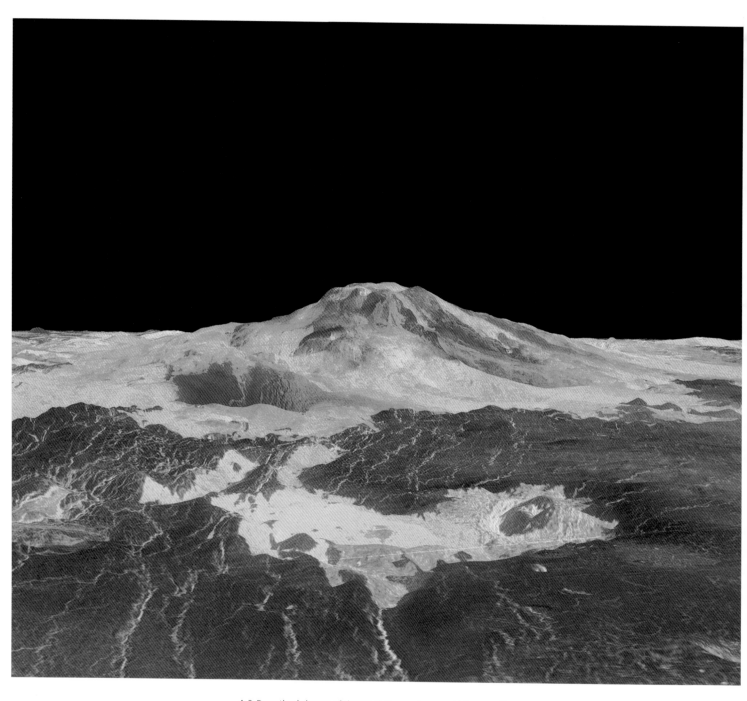

A 3-D synthesis image of the Maat Mons volcano, which tops off at 5,000 meters above the neighboring lava plains.

FERDINAND MAGELLAN

(1480-1521)

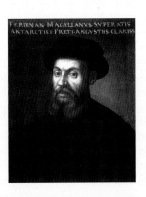

Ferdinand Magellan (Fernão of Magalhães) was a Portuguese sailor and explorer. He is renowned for organizing and commanding the first circumnavigation of the Earth, between 1519 and 1522. This voyage was sponsored by the Spanish Crown, who wanted to find a new western spice route between the Maluku Islands (known then as the Spice Islands, where cloves and nutmeg were originally cultivated) and Europe. The expedition, which began with five ships, lasted three troublesome years and survived a mutiny, an episode of scurvy, a drowning and a hostile boarding party. Magellan died in 1521 in a battle waged against natives on the isle of Mactan in the Philippines. He discovered the strait between the Atlantic and Pacific Oceans at the tip of South America, which was named in his honor. It would be 58 years before Sir Francis Drake would complete the second successful circumnavigation of the world, in 1580. The route to the Pacific via the Strait of Magellan was not used for several centuries, because of the treacherous navigation conditions involved.

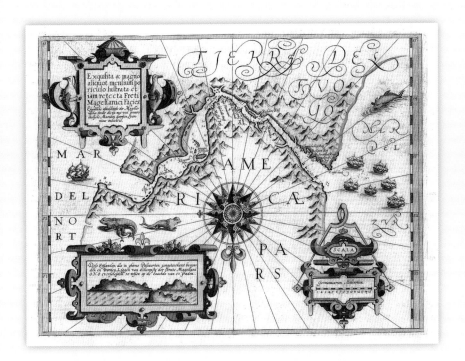

Venus Express

V ENUS EXPRESS is the first Venus exploration mission attempted by the European Space Agency (ESA). This mission was proposed in 2001 and is based on the previously developed Mars Express probe (see page 184), with which it shares many characteristics. A few structural changes had to be incorporated for this mission, taking into account the very different environments on the two planets. More thermal protection and more protection against ionizing radiation were needed for the Venus trip. On the other hand, since sunlight is four times stronger near Venus than near Mars, the same photovoltaic solar panels furnished more energy to this probe. The main goals of the mission were a detailed study of the physical and chemical characteristics of the Venusian atmosphere and clouds, and the mapping of surface temperatures. The Venus Express digital camera, with its multiple channels, was modeled after the high-resolution cameras used by the Mars Express and Rosetta (see page 313)

Artistic rendition of Venus Express in working orbit.

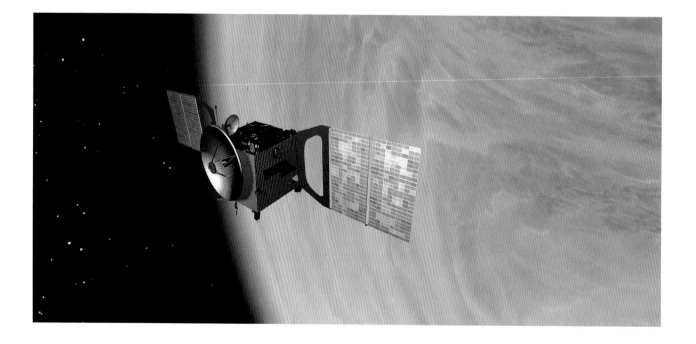

EADS ASTRIUM

Raising the Soyuz-Fregat rocket into firing position, with the Venus Express on board, at the Russian Baikonur Cosmodrome in Kazakhstan. With more than 1,700 launches to its credit, Soyuz is the most utilized rocket in the world.

LEFT

Assembly of the Venus Express in the EADS (European Aeronautic Defense and Space Company) Astrium site in Toulouse, France.

BELOW

Installation of the Venus Express orbital probe into the Fregat enclosure.

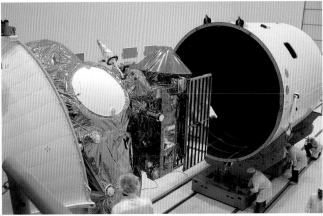

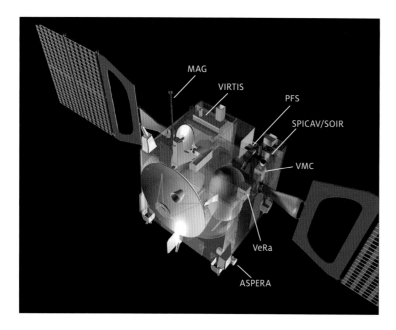

Venus Express carries a battery of scientific instruments to study the planet, including a magnetometer (MAG), a spectrophotometric imager (VIRTIS), two spectrometers (PFS and SPICAV/SOIR), a camera (VMC), the Venus Express Radio Science Experiment (VeRa), and a plasma analyzer (ASPERA).

RIGHT

Two faces of Venus. This image represents the nighttime view, as seen through infrared (red-orange) imaging, and the daytime view, seen through ultraviolet light (blue) imaging, provided by the VIRTIS spectrometer of Venus Express. The infrared radiation locates the deep cloud layers, while ultraviolet radiation indicates the distribution of clouds in the upper atmosphere of Venus.

probes. A digital processor was added to the imaging system to compress the raw data before sending them, so as to save on bandwidth for radio communications with Earth. Several instruments developed for other ESA missions were used on the Venus Express, including the PSF (Planet Fourier Spectrometer), the SPICAV/SOIR (Spectroscopy for the Investigation of Characteristics of the Atmosphere of Venus/ Solar Occultation at Infrared) spectro-meters, the radio probe VeRa and the MAG magnetometer. Other scientific equipment on board were the ASPERA (Analyzer of Space Plasmas and Energetic Neutral Atoms) and the imaging VMC (Venus Monitoring Camera). Venus Express lifted off from Baikonur on November 5, 2005, propelled by the Soviet Soyuz-Fregat rocket into stationary Earth orbit. After that, the probe's Fregat rocket booster was fired, sending Venus Express on a 153-day trajectory to Venus. On April 11, 2006, the spacecraft was captured by the planet's gravitational field. Several maneuvers were required to place it in its final working orbit: a 24-hour elliptical polar orbit of 249 by 66,582 kilometers, which was attained on May 18, 2006. The probe

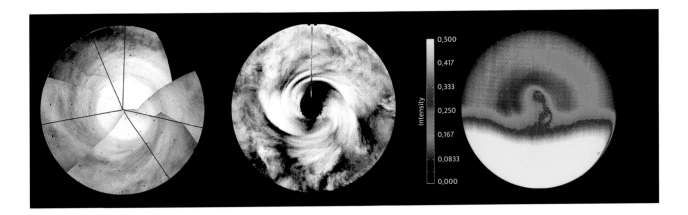

transmitted its first images on April 13. The results suggest that oceans were present on the planet in the past. Observations of the South Pole with the VIRTIS (Visible and Infrared Thermal Imaging Spectrometer) led to the discovery of a huge double atmospheric vortex, reminiscent of Earth's cyclones. The detection of sulfur dioxide in the atmosphere raises the question of whether there are active volcanoes on Venus. The Venus Express data also confirmed that lightning was more frequent on Venus than on Earth. Having perfectly completed its nominal phase in the course of two Venusian days (500 Earth days), the Venus Express mission is currently extended until December 2012. ♀

Images of the South Pole of Venus taken by three different orbital probes: Mariner 10, Pioneer Venus Orbiter and Venus Express. Only the precise images by Venus Express revealed a double atmospheric vortex.

 # Is there life on Earth?

Living organisms modify the atmospheric composition of our planet. To test new long-distance detection methods for potential life on exoplanets (those outside the solar system), Venus Express pointed its VIRTIS spectrometric camera toward Earth several times in 2007, from its vantage point orbiting Venus. The analysis of light reflected from Earth allowed it to identify spectral signatures of several molecules present in the atmosphere, including carbon dioxide (CO_2), methane (CH_4), ozone (O_3) and nitrous oxide (N_2O). During this experiment, no surface detail of Earth, 46 million kilometers away, was visible, not surprisingly since our planet occupied only one pixel in that entire image.

Mars: The Red Planet

MARS IS THE FOURTH PLANET from the Sun. After the Sun, the Moon and Venus, it is the brightest object in the sky (when closest to Earth). It has been nicknamed "The Red Planet," because when viewed with the naked eye it has an orange-ocher color, caused by its surface, which is rich in iron-oxide. Its atmosphere is composed primarily of carbon dioxide.

Mars is a terrestrial planet with very prominent volcanic uplifting. It is home to the largest volcano in the solar system, Olympus Mons, which rises more than 20 kilometers above the surrounding plains. Mars is also home to the largest canyon known, Valles Marineris, which is 4,000 kilometers long and 7 kilometers deep. In contrast, the Grand Canyon in Arizona is just 446 kilometers long and 2 kilometers deep!

Our notions about habitable planets (i.e., planets capable of sustaining life) assume the presence of liquid water. Of all the planets studied, Mars is the only one, along with the Earth, whose geologic record indicates that large quantities of liquid surface water could have been present there in the past. Mars is consequently an essential planet in our understanding of how life began.

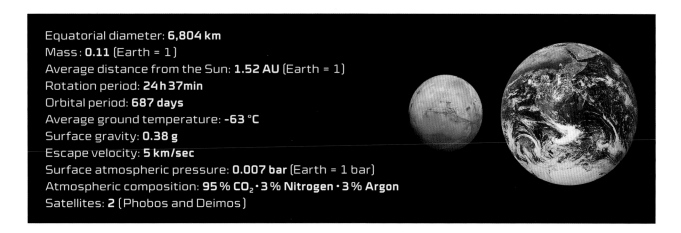

Equatorial diameter: **6,804 km**
Mass: **0.11** (Earth = 1)
Average distance from the Sun: **1.52 AU** (Earth = 1)
Rotation period: **24 h 37 min**
Orbital period: **687 days**
Average ground temperature: **-63 °C**
Surface gravity: **0.38 g**
Escape velocity: **5 km/sec**
Surface atmospheric pressure: **0.007 bar** (Earth = 1 bar)
Atmospheric composition: **95 % CO₂ · 3 % Nitrogen · 3 % Argon**
Satellites: **2** (Phobos and Deimos)

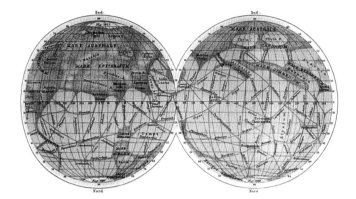

A little history

Since the color of Mars is reminiscent of blood, it was named Nergal — the god of war, fire and destruction — by the ancient Babylonians. The Romans in turn named it Stella Martis — the star of Mars — to honor their god of war.

At the end of the 19th century, Giovanni Schiaparelli, an Italian astronomer and director of the Milan observatory, drew an accurate map of the surface of Mars. He also first described what he thought were irrigation canals (*canali* in Italian) conveying water from the frozen poles to the equator.

In 1600, German mathematician Johannes Kepler was hired by the celebrated Danish astronomer Tycho Brahe (who had just finished building his observatory near Prague), to determine the orbit of Mars. Kepler discovered that the planets moved in elliptical orbits, not circles, around the Sun; thus he formulated his First Law of planetary motion. Dutch astronomer Christian Huygens (1629–1685, see page 253) observed flat, dark features on Mars suggestive of seas. William Herschel (1738–1822), the discoverer of Uranus, was also captivated by the red planet, and was convinced that Mars was very similar to the Earth by virtue of climate and geography. Huygens and Herschel were among the first to envision that the planet Mars could be inhabited.

At the end of the 19th century, Italian astronomer Giovanni Schiaparelli (1835–1910) mapped the surface of Mars very accurately and described what he thought was a system of irrigation canals (*canali*), which conveyed the frozen polar waters to the arid equatorial regions. For his part, Percival Lowell (1855–1916), in his observatory in Flagstaff, Arizona, confirmed the presence of canals and imagined abundant vegetation and a breathable atmosphere. More rigorous observations by Edward E. Barnard (1857–1923), however, refuted the presence of seas and canals on Mars as nothing more than optical illusions fed by the fertile imaginations of Schiaparelli and Lowell.

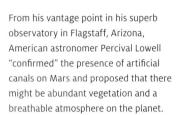

From his vantage point in his superb observatory in Flagstaff, Arizona, American astronomer Percival Lowell "confirmed" the presence of artificial canals on Mars and proposed that there might be abundant vegetation and a breathable atmosphere on the planet.

The red planet has two natural satellites. These moons were discovered by Asaph Hall in 1877, who named them Phobos (Greek for "fear") and Deimos ("terror"), after the two sons of Ares, the ancient Greek god of war.

Mars and science fiction

The idea of an advanced extraterrestrial civilization capable of building a network of irrigation canals on a planetary scale has given birth to science fiction themes in popular culture. The first great classic science fiction novel, *War of the Worlds*, by H.G. Wells (1897) tells of an invasion of the Earth by Martians. When adapted as a supposed radio report by Orson Wells in 1938, the story created widespread panic on the east coast of the United States. Many movie versions of this story have been made: the two most famous are the 1953 Byron Haskin film and the 2005 Steven Spielberg adaptation, both greatly enjoyed by fans of dramas with high emotional content.

Mars and space probes

Since the 1960s, several dozen space probes, both orbiters and landers, have been sent on exploratory missions to Mars by the Soviet Union, the United States, Europe and Japan. A large number of these missions failed, due to loss of radio contact, crashing or other mysterious technical reasons, which gave rise to many strange bad rumors in the astronautical community. To counter these rumors, it must be stressed that a successful mission to Mars requires that a number of complex factors work in unison. After launch, the spacecraft is injected into a Hohmann transfer orbit (see page 19): an orbit whose perihelion, the Earth, is the departure point and whose aphelion, the arrival point, is Mars. As a result, the launch must be timed precisely to the day, and even the hour, relative to the position of Earth and Mars in their respective orbits, to insure that Mars is exactly at the aphelion point of its orbit when the probe arrives. The Soviet Union attempted to launch a probe to the red planet five times between October 1960 and November 1962. Only Mars 1, an 890-kilogram probe launched on November 1, 1962, succeeded in escaping Earth orbit. Unfortunately its attitude adjustment system failed and communications with the spacecraft was lost when it was 106 million kilometers from the Earth. Another two years passed before the first successful U.S. Mars probe, Mariner 4. ♂

Martian attack, illustrated by Alvim Correa for the novel *War of the Worlds*.

1964 Mariner 4

Successful launch of Mariner 4 by an Atlas-Agena rocket toward Mars.

RIGHT

Last-minute preparations before launch.

BASED ON THE SAME octagonal design as Mariner 2, the two Mars probes Mariner 3 and 4 were designed primarily to fly over the planet and capture the first close-up images of its surface. Carried by an Atlas-Agena D rocket, Mariner 3 was launched November 5, 1964, but was unable to reach its target because it could not jettison its protective cover. However, its twin, Mariner 4, was successfully launched from the Kennedy Space Center at Cape Canaveral on November 28, 1964. It was able to jettison its shroud and then headed toward Mars. After a more than 7-month interplanetary trek without incident, Mariner 4 flew by Mars on July 14 and 15, 1967, and transmitted 22 historic TV images of the planet's surface. The spacecraft came to within 9,846 kilometers of the planet. Communications with it ceased on December 21, 1967, but both Mariner 3 and 4 still orbit the Sun. In the interim, the Soviet Union launched Zond 2 on November 30, 1964. Skimming past Mars at 1,609 kilometers on August 6, 1965, the probe was unable to transmit the planned photos, due to a glitch in the communications system. Zond 2 was the first spacecraft to utilize an ion engine for attitude control. ♂

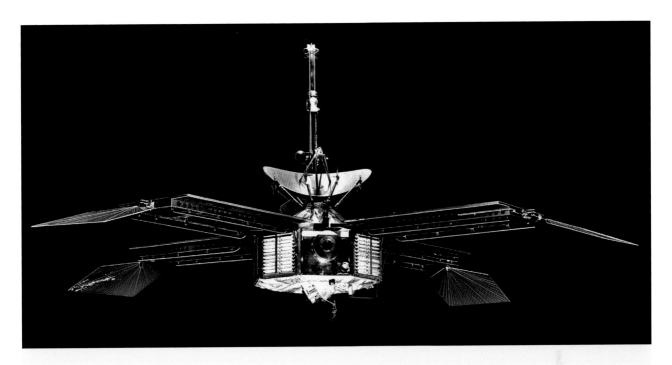

MAIN COMPONENTS OF MARINER 4

1. Low-gain antenna
2. Magnetometer
3. Particle-flux detector
4. Solar panels
5. Thermal-control vents
6. Star tracker
7. Sun tracker
8. Television camera
9. Infrared spectrometer
10. Movable solar panel
11. Propulsion nozzle
12. High-gain antenna

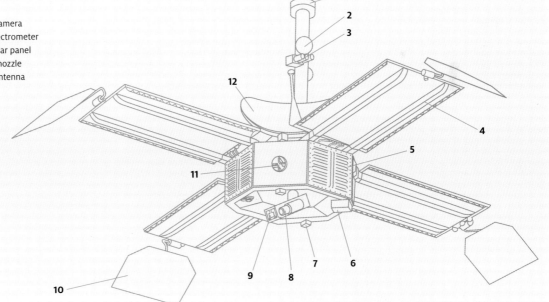

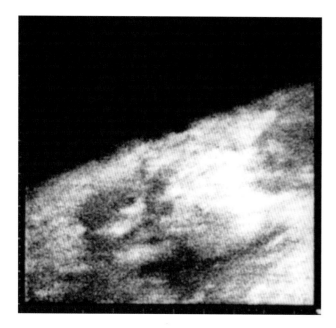

First image of Mars sent by a space probe.

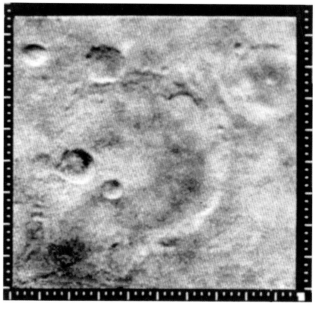

First evidence of craters on Mars.

Anxious to see images transmitted from Mars before they were computer processed, the telecommunications technicians manually decoded the range of numbers in real time by composing a sort of paint-by-numbers picture.

The vidicon-type of television camera sent aboard Mariner 4.

NASA and JPL directors present U.S. President Lyndon Johnson with Mariner 4 images of Mars, during an official ceremony in 1965.

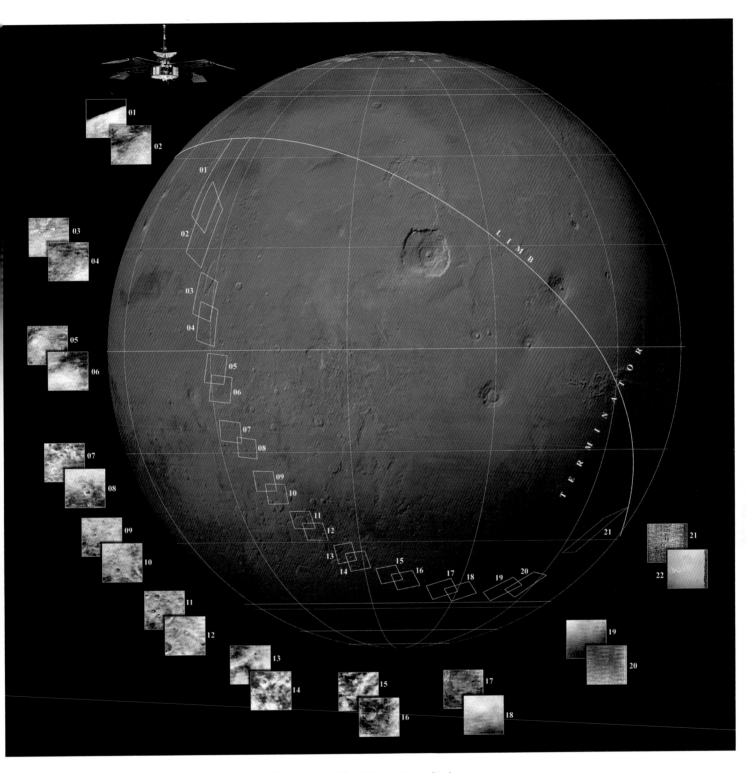

Trajectory of the Mariner 4 flyby and geographical position of the 22 images transmitted.

Mariner 6 and 7

N FEBRUARY 24, 1969, the Mariner 6 rose from launch complex 36B at Cape Canaveral in Florida, beginning a five-month journey to Mars. Launched by the newly deployed Atlas-Centaur rocket, the Mariner 6 mission was primarily to transmit images of the Martian surface and to obtain data on its atmosphere during a brief flyby, in advance of its twin probe, Mariner 7 (launched March 27, 1969). The launch of Mariner 6 came close to disaster. Due to a valve failure, the 12-storey high rocket began to crumble. At the risk of their lives, two NASA technicians were able to intervene and save its precious cargo for mounting on another rocket. The Mariner 7 mission also avoided disaster when one of the probe's batteries exploded a few days before the scheduled flyby, resulting in a loss of radio contact. Fortunately, contact was re-established with just a few hours to spare.

Launch of the Atlas-Centaur rocket carrying Mariner 7, on March 17, 1969.

OPPOSITE PAGE, TOP

Preparation of the Mariner 6 probe. Technicians are carrying out a balance test to determine the spacecraft's center of gravity.

OPPOSITE PAGE, BOTTOM

The Mariner 6 space probe.

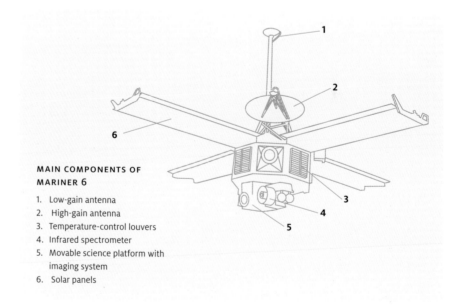

MAIN COMPONENTS OF MARINER 6

1. Low-gain antenna
2. High-gain antenna
3. Temperature-control louvers
4. Infrared spectrometer
5. Movable science platform with imaging system
6. Solar panels

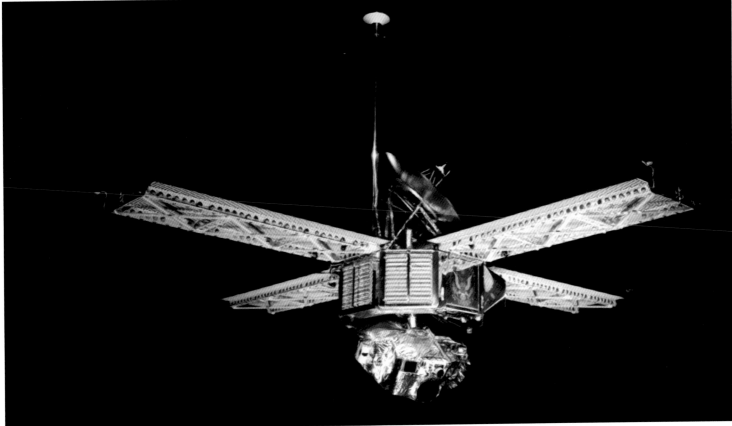

Collage of the wide-angle images obtained by Mariner 6 and 7 on Mars, as seen from Earth. Mariner 6 images are the two horizontal strips over the northern hemisphere, and the Mariner 7 images are the two diagonal strips over the southern hemisphere.

TOP RIGHT

In 1967, computers were not as sophisticated as today. During preparation for the Mariner 6 and 7 missions, NASA technician used a turntable to simulate Martian rotation and to program the operating times of the cameras and different instruments during the flyby.

In July of 1969, as the two Mariner spacecraft were approaching their target, the whole world was focused on the Apollo 11 mission. In fact, the first steps on the Moon, when Neil Armstrong said his celebrated "That's one small step for a man, one giant leap for mankind," took place on July 20, 1969. The Mars flybys by Mariner 6 and 7 were therefore almost eclipsed by this historic event, despite the undeniable success of this double mission. Each probe was equipped with two television cameras, one with a 50 mm widefield lens and the other with a longer focal length (500 mm) lens for more detailed views. In addition, each probe carried a number of other scientific instruments: an infrared radiometer to measure the surface temperature of Mars, an ultraviolet spectrometer to study the composition of the upper atmosphere and an infrared spectrometer to study the composition of the lower atmosphere. Thanks to the spectacular advances in electronics over just a few years, the observing and communications capacity of this new generation of Mariners far exceeded those of their predecessors. It took Mariner 4 eight hours to transmit a single image of mediocre quality (200 x 200 pixels, 64 gray levels) to Earth, while Mariner 6 required only 5 minutes to transmit a much superior image (800 x 800 pixels, 128 gray levels). Moreover, thanks to a new type of on-board computer that could be programmed in flight, engineers were able to modify the trajectory of Mariner 7 and compensate for the effects of the exploded battery on the imaging system, while monitoring information from Mariner 6.

More than 200 video images of the Martian surface were taken (76 by Mariner 6, 159 by Mariner 7), primarily of equatorial regions and the southern hemisphere, all from a minimal altitude of between 3,330 and 3,518 kilometers. The two spacecraft were also used to test Einstein's theory of relativity by measuring (with an accuracy of five nanoseconds) the delay to which an electromagnetic signal was subjected, when passing close to a massive object like the Sun.

The images transmitted by Mariner 6 and 7 confirmed the absence of Lowell's canals on Mars. As indicated earlier by Mariner 4, certain portions of Mars resemble the Moon with its numerous craters. Nonetheless, with its varied relief, its atmosphere and the presence of an ice cap at the South Pole composed of frozen carbon dioxide, Mars was different from all known celestial objects. The next technical step would be to go beyond a quick flyby in favor of an extended orbital tour of Mars, to understand more about it. The northern hemisphere would prove particularly surprising. ♂

Layout of the science platform and imaging system:
A Infrared radiometer
B Wide-angle TV camera
C Ultraviolet spectrometer
D Close-up TV camera
E Infrared spectrometer

Global images of Mars by Mariner 7, taken during the approach phase.

Mariner 9

BELOW, LEFT
Technicians ready Mariner 9 for encapsulation prior to launch.

BELOW, RIGHT
Mariner 8 did not reach its destination, due to problems during launch, but Mariner 9 was successfully launched toward Mars on May 30, 1971, from Cape Canaveral.

OPPOSITE PAGE
The Mariner spacecraft.

The 1971 launch window saw a flotilla of orbiters sent to Mars: three Russian and two American. During this period of international Cold War rivalries, only one mission was able to achieve its goal, amid multiple difficulties. On May 8, the American craft Mariner 8 crashed into the Atlantic Ocean only a few minutes after its launch from Cape Canaveral. The upper stage of the Atlas-Centaur rocket deviated dangerously from its trajectory, due to a guidance system malfunction and the decision was made to destroy it for safety reasons. This was a really bad start for the most ambitious American planetary exploration program ever attempted. The

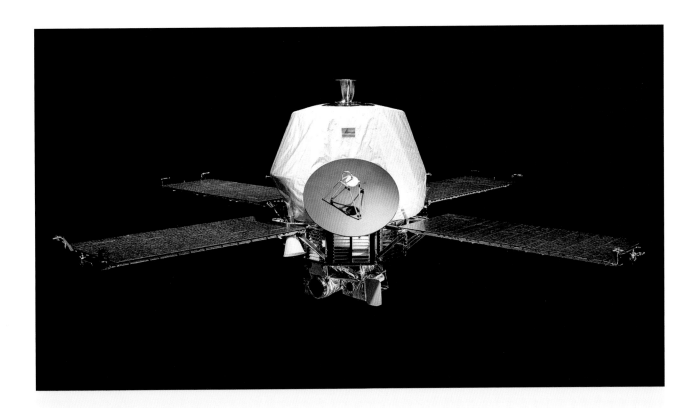

MAIN COMPONENTS OF MARINER 9

1. Low-gain antenna
2. Propulsion engine
3. Propellant tank
4. Star tracker
5. Pressurization tank
6. Solar panel
7. Long focal-length TV camera
8. UV spectrometer
9. Widefield TV camera
10. Infrared radiometer
11. Medium-gain antenna
12. High-gain antenna

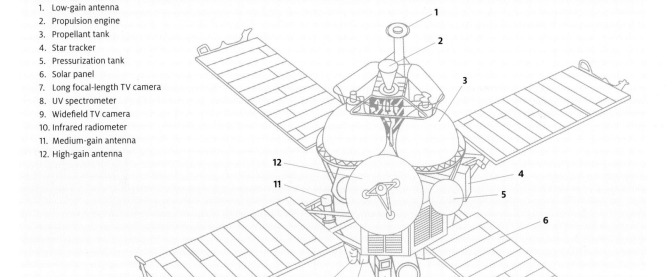

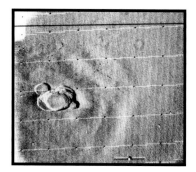

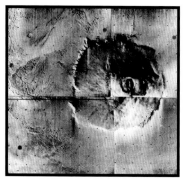

TOP
A global dust storm did not reveal anything to the orbiting Mariner 9 except the summit of the volcano Olympus Mons.

BOTTOM
A few weeks later, after the dust settled, the solar system's largest volcano, Olympus Mons, was revealed in all its glory.

Mariner 8 and 9 missions were designed to complement one another in their exploration of Mars. The 22-day delay between their two launches was hectic for scientists and engineers, who had to completely reprogram Mariner 9's mission. When Mariner 9 left the Earth on May 30 without incident, NASA could breathe easier. On November 13, 1971, after a journey of six months, it entered Mars orbit as planned, at an altitude of 1,390 kilometers, becoming the first artificial satellite to circle another planet, beating Mars 2 by a few days. It first images, however, were quite deceptive; Mars looked like a grey ball without detail or relief features, with four unrecognizable dark patches! As the final straw, Mars was enveloped by a planet-wide dust storm. This phenomenon had long been observed by astronomers, who had actually predicted a storm at around this time, but no one in charge of the mission had consulted them. There was nothing to do but wait. More than a month passed before the dust settled and Mars finally revealed its secrets.

The dark patches, as it turned out, were the tops of three giant volcanoes in the Tharsis region of Mars, topped by the summit of the immense volcano Olympus Mons, whose caldera reaches a height of 22 kilometers (measured by laser altimetry), making it the tallest mountain in the solar system. One of the most important discoveries by Mariner 9 was the presence of numerous fluvial valleys, including the giant canyon Valles Marineris, clearly formed during a period of extensive water flow on Mars.

Over the course of a year of steady and reliable service, Mariner 9 captured 7,329 images, mapping the entire Martian surface, as well as gathering much data on the planet's atmosphere, geology and meteorology. The spacecraft also sent close-up photos of the two Martian moons, Deimos and Phobos. In the annals of space exploration, this mission counts as an outstanding success. Thanks to Mariner 9, Mars was again seen as an active planet of great scientific interest. Above all, the evidence that erosion due to water was a major factor on Mars once again raised hope of finding life there, in the present or in fossil form. The subsequent American Viking mission and landing was focused on that fascinating question. ♂

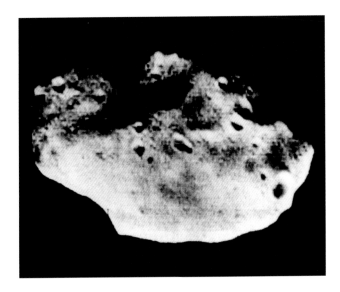

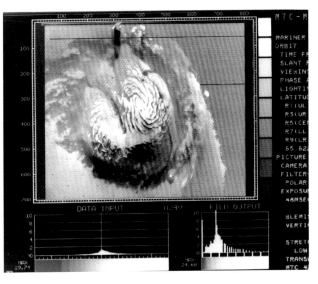

The first image of the moon Phobos, taken by Mariner 9 from a distance of 5,760 kilometers. The moon is about 26 kilometers long, and many surface craters are evident.

The north polar cap during the Martian summer.

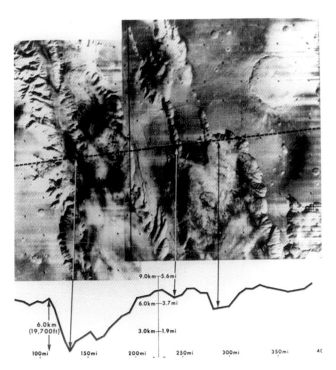

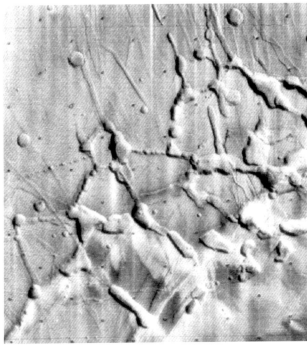

The giant canyon that crosses the northern hemisphere of Mars was named Valles Marineris in honor of Mariner 9, which discovered it.

The Noctis Labyrinthus region, west of Valles Marineris, is covered by a network of canyons.

Mars 2 and 3

One of the few images of Mars sent by Mars 2, as it approached the planet during the dust storm of 1971.

OPPOSITE PAGE
Scale model of the Mars 2 space probe.

THE SOVIET UNION planned to fully exploit the favorable Mars window of 1971 with no less than three probes — Kosmos 419, Mars 2 and Mars 3 — all launched by Proton rockets from Baikonur. Launched May 10, Kosmos 419 ended up in an Earth parking orbit, due to a programming error in firing the booster rocket, and fell back to Earth, burning up like a meteor. The twin probes Mars 2 and 3, each weighing 4,644 kilograms, were successfully launched on May 19 and 28, respectively. Each spacecraft consisted of two independent components: an orbiting platform equipped with a television camera and a panoply of scientific instruments, and, for the first time, a lander, also equipped with a camera to image the surface and several instruments to study the soil. They each also carried a mobile mini-robot. No problems were encountered during the voyage, and both craft were successfully placed into Martian orbit (November 27 for Mars 2 and December 2 for Mars 3). Unfortunately both spacecraft were also hindered by the same dust storm that affected Mariner 9. Unlike Mariner, however, the Mars 2 and 3 missions were programmed to run automatically, with no possibility of change. This lack of flexibility proved catastrophic. The landers were consequently deployed during the storm into winds with speed up to 400 kilometers per hour. The Mars 2 lander crashed and the Mars 3 lander only lasted two minutes. So, although Mars 3 was the first human-made object to land on the red planet, it did not transmit any usable information. The orbiters continued to function for several weeks, thereby avoiding the humiliation of total failure. Despite the meager scientific returns and the failure of the landers, Soviet engineers were nonetheless satisfied to have reached Mars. Refusing to give up in the face of their bad luck, they were already preparing for the 1973 window of opportunity. ♂

Mars 5

The americans did not send a second space probe to Mars in 1973, focusing instead on readying the Viking mission for the more favorable window in 1975. During that time at Baikonur, no less than four Soviet probes were readied for the Proton rockets, including Mars 4 to 7, for launch during the summer of 1973. This launch window was less favorable than the previous ones. In effect, more fuel would be needed to attain the proper trajectory, to the detriment of the scientific payload. Consequently, the Soviets could not combine orbiters and landers in the same launcher, as was done previously.

Full-scale model of Mars 5.

The Mars 4 and 5 missions, launched July 21 and 25, contained only the orbiter components. They were intended primarily for planetary imaging and to relay signals from the landers, Mars 6 and 7, launched August 5 and 9. But all of this promising operation had a hidden flaw. In trying to explain what had gone wrong during a prelaunch test, officials learned with dismay that two years before, as a cost-saving measure for the country's declining reserves, an over-zealous bureaucrat had ordered a change from gold to aluminum transistors. Unlike gold, aluminum is subject to corrosion, and analysis of the inferior transistors showed that their reliability decreased markedly two years after manufacture — in other words, right during the Mars mission! Lacking time to replace all suspect electronic components on all the spacecraft before launch, and estimating the chance of equipment survival at 50 percent, the decision to continue was made without illusion.

Mars 4 arrived at its destination on February 10, 1974, but its retrorockets stopped at 2,100 kilometers from the surface and the probe was lost in space, in solar orbit. Mars 5, the only operational craft of the mission, was placed in proper orbit on February 12, 1974, and returned data and 108 color photos of Mars. Unfortunately, two weeks later, contact with Earth was permanently lost. Mars 6 lost communication by telemetry two months into its trip. The on-board lander was deployed as planned and its radio signals detected, but 148 seconds after deployment of its parachutes, it went silent before reaching the surface. Nevertheless, the initial bits of data on the chemical composition of the Martian atmosphere were transmitted. The doomed mission continued. Mars 7 arrived at its destination on March 9, 1974, and launched its lander prematurely into empty space, probably due to a defective transistor.

This mission will remain as a great failure in the annals of space exploration. Still, it must be noted that a spirit of cooperation existed between American and Soviet planetologists. Together, they analyzed Mars 5 images and compared them to those of Mariner 9, thereby managing to produce the first accurate surface maps of Mars. Fifteen years would pass before the Soviets would again take their chances with the red planet. ♂

The two types of photo-television cameras on board Mars 5, similar in design to those on the lunar probe Zond 3.

BELOW

Images of the Martian surface recorded by Mars 5, with a 350 mm focal length lens (left) and the 52 mm widefield view (right).

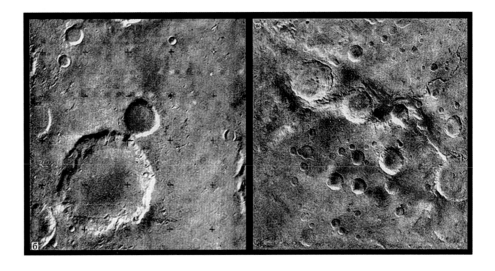

1975

The Viking Mission

BELOW, LEFT

Last preparations before takeoff: the nose cone of the Titan IIIE-Centaur rocket is carefully positioned on the Viking 1 spacecraft with its solar panels folded in launch configuration.

BELOW, RIGHT

Launch of Viking 1 from Cape Canaveral on August 20, 1975.

OPPOSITE PAGE, TOP

The Viking probe in its cruising configuration. The protective cocoon housing the lander is attached to the orbiter body, shown with its solar panels extended.

Encouraged by the success of Mariner 9, NASA undertook an ambitious and detailed study of Mars with the Viking mission. The mission included two probes, Viking 1 and Viking 2, each consisting of two vehicles, an orbiter and a lander. Each spacecraft had a combined weight of 3,521 kilograms: the orbiter, based on the Mariner design, weighed 2,327 kilograms and the descending stage, 1,194 kilograms. The primary objectives were to obtain high-resolution images of the planet, to characterize the structure and composition of the surface and atmosphere, and, for the first time, to search for evidence of extraterrestrial life. The technical success and amount of new data accumulated about Mars renders this spectacular undertaking one of the highlights of the history of solar-system exploration.

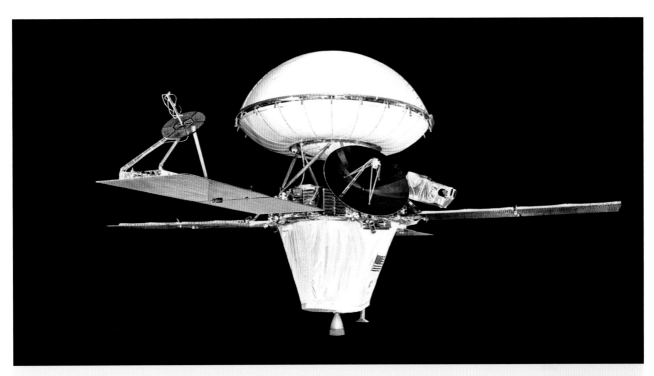

MAIN COMPONENTS OF A VIKING ORBITER

1. Aerodynamic shield
2. Biological capsule
3. High-gain antenna
4. Thermal-control unit
5. Solar panel
6. Star tracker
7. Propulsion module
8. Engine nozzle
9. Low-gain antenna
10. Pointable scan platform

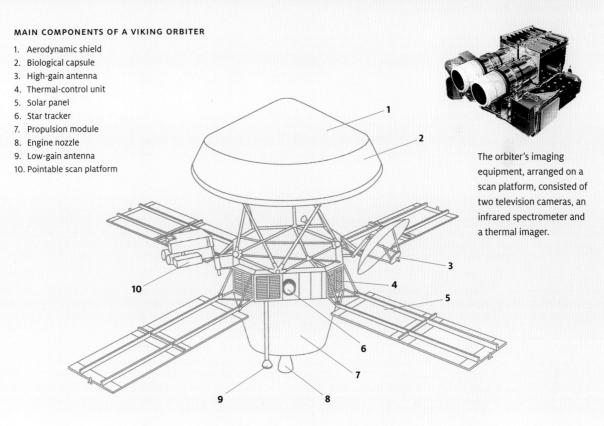

The orbiter's imaging equipment, arranged on a scan platform, consisted of two television cameras, an infrared spectrometer and a thermal imager.

The sequence of entry, aerobraking and landing of the Viking landers.

The Viking 1 and 2 were launched from Cape Canaveral by the powerful Titan IIIE-Centaur rockets on August 20 and September 9, 1975, respectively. After an interplanetary voyage of a little less than one year, Viking 1 arrived at its destination on June 19 and Viking 2 on August 7, 1976. The first month in orbit was used to find the best landing sites, both the safest and most scientifically interesting. The Viking 1 lander was deployed on July 20, 1976, and safely set down in Chryse Planitia region. The Viking 2 lander set down on Utopia Planitia on September 3, 1976. After more than 700 orbits, the Viking 2 orbiter ceased communications on July 25, 1978, due to a command error. For its part, the Viking 1 orbiter was "unplugged" on August 17, 1980 after more than 1,400 orbits. Together, the orbiters transmitted more than 50,000 images that cover almost the entire planetary surface. Initially designed for 90 days of operation on the surface of Mars, the two landers continued to function for several years. The Viking 2 lander went silent on April 11, 1980, and Viking 1 on Sept 13, 1982. ♂

TRAJECTORIES OF THE INTERPLANETARY VOYAGES OF THE SPACECRAFT VIKING 1 AND VIKING 2

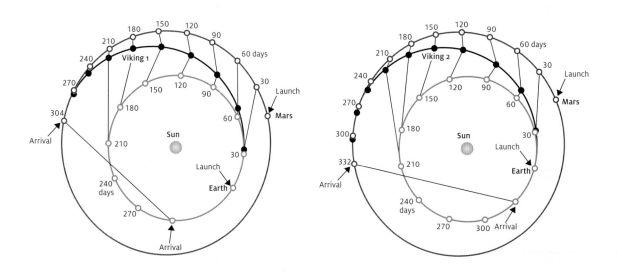

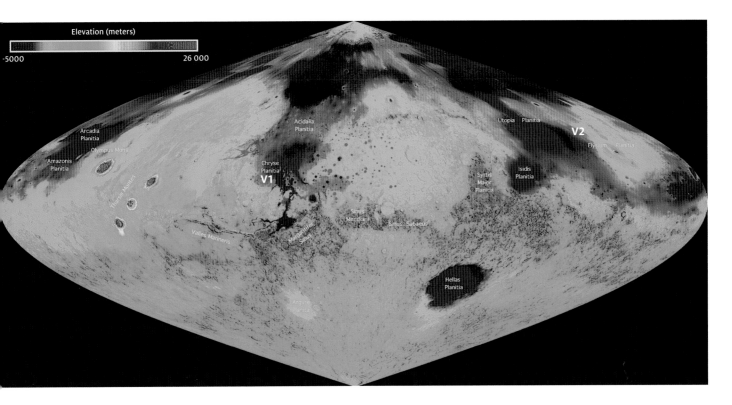

Elevation (meters)

-5000   26 000

Arcadia
Planitia

Olympus Mons

Amazonis
Planitia

Tharsis Montes

Valles Marineris

Acidalia
Planitia

Chryse
Planitia
V1

Margaritifer
Sinus

Sinus
Meridiani

Sinus Sabaeus

Argyre
Planitia

Hellas
Planitia

Utopia   Planitia
V2

Elysium   Planitia

Isidis
Planitia

Syrtis
Major
Planitia

Geographic location of the Viking mission landers; Viking 1 on the plain Chryse Planitia, close to the equator and Viking 2 on the plain Utopia Planitia, closer to the North Pole.

JPL technicians inspect the heat shield, whose critical role is to protect the lander from the intense heat caused by the high-speed entry of the craft into the Martian atmosphere.

LEFT

A protective film is placed over the Viking lander to prevent biological contamination.

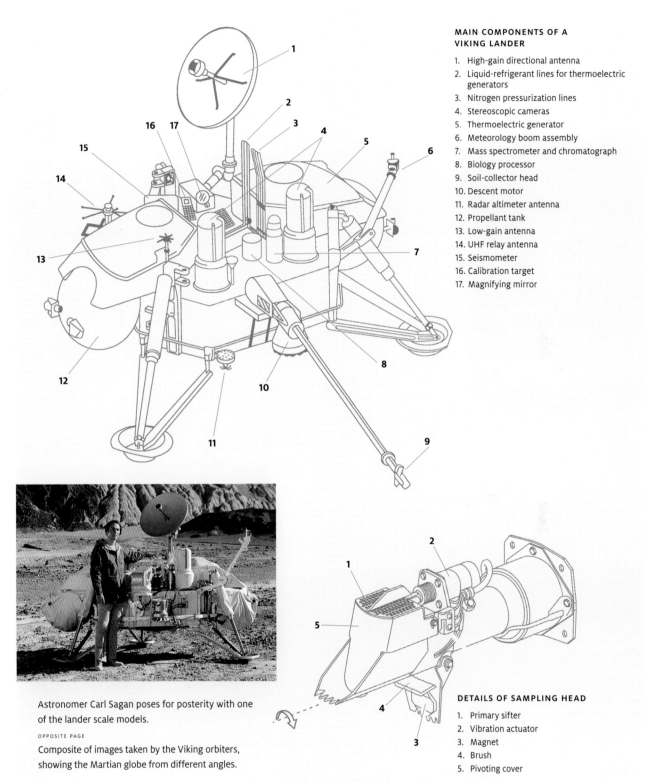

MAIN COMPONENTS OF A
VIKING LANDER

1. High-gain directional antenna
2. Liquid-refrigerant lines for thermoelectric
   generators
3. Nitrogen pressurization lines
4. Stereoscopic cameras
5. Thermoelectric generator
6. Meteorology boom assembly
7. Mass spectrometer and chromatograph
8. Biology processor
9. Soil-collector head
10. Descent motor
11. Radar altimeter antenna
12. Propellant tank
13. Low-gain antenna
14. UHF relay antenna
15. Seismometer
16. Calibration target
17. Magnifying mirror

Astronomer Carl Sagan poses for posterity with one
of the lander scale models.

OPPOSITE PAGE

Composite of images taken by the Viking orbiters,
showing the Martian globe from different angles.

DETAILS OF SAMPLING HEAD

1. Primary sifter
2. Vibration actuator
3. Magnet
4. Brush
5. Pivoting cover

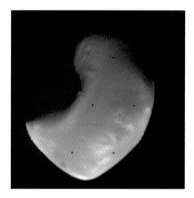

Zooming in on the two Martian moons:
Phobos (upper) and Deimos (lower),
photographed by the Viking orbiters.

A face on Mars

This image, taken by the Viking 1 orbiter on July 25, 1976, in the Cydonia region, has caused much ink to flow. Some commentators saw an artificial human face there and proof of a long-gone Martian civilization. NASA decided to re-photograph this strange feature in 2001 with the Mars Global Surveyor and its high-resolution camera (see page 163). The Mars Express probe also mapped this feature in 3-D.

A historic document: The first image of the surface of Mars and the first image taken on the surface of another planet by an American spacecraft. The Viking landers used digital CCD (Charge-Coupled Device) imagers, an advanced technology at that time. The inventor of the CCD camera, Canadian physicist and Nobel laureate Willard Boyle, declared that this image of Mars counts among the greatest successes of his career.

CARL SAGAN

(1934–1996)

Born in Brooklyn, New York, physicist and American science popularizer Carl Sagan was professor of astronomy and space sciences at Cornell University. Multidisciplinary researcher and prolific author, he published more than 600 articles and more than 20 books, some of which were world-wide bestsellers (*Other Worlds, Cosmos, Shadows of Forgotten Ancestors, The Demon-Haunted World*). As a pioneer in exobiology (the search for extraterrestrial life), he participated in the establishment of SETI (Search for Extra-Terrestrial Intelligence) and denounced the dangers of a post-nuclear winter. The public at large also knew him as the presenter of the PBS television series Cosmos. As a NASA consultant, Carl Sagan contributed to numerous solar-system exploratory missions, including Pioneer 10 and 11, Mariner 9, Viking, Voyager and Galileo.

"Imagination will often carry us to worlds that never were. But without it we go nowhere."

A panoramic view of Utopia Planita taken by Viking 2. Color was obtained by combining images of the scene taken through red, green and blue filters.

Traces of life on Mars?

It's widely agreed in the scientific community that life on Mars could only exist in the form of microorganisms. The Viking lander cameras did not detect any form of life: animal nor plant. Four approaches were taken for the first experiment in exobiology for detecting microorganisms on Mars.

1. The presence of organic molecules

All living *organisms* on Earth are composed of organic molecules. A gas chromatograph and mass spectrometer were used to look for organic molecules in the Martian soil as evidence of life, past or present. No traces of such compounds were detected. This equipment was so sensitive that it detected traces of solvents used during assembly of the spacecraft.

2. Pyrolytic release experiment

Living organisms on Earth fix carbon from atmospheric CO_2 in order to synthesize more complex organic molecules. This experiment was based on radioactive ^{14}C fixation, which was provided on board as $^{14}CO_2$ gas for assimilation either by photosynthesis or metabolic intake. After five days of incubation in the presence of $^{14}CO_2$, the sample was heated to 650°C to vaporize any molecules tagged with ^{14}C. The Geiger counter did not detect any carbon fixation.

3. The labeled release experiment

A soup of nutrient molecules labeled with ^{14}C was added to a soil sample to look for the release of labeled carbon dioxide, generated by metabolic activity of microorganisms. Surprisingly, ^{14}C labeled gas was effectively released and heat sterilization of the sample reduced such activity markedly. These results were interpreted as an argument in favor of heat-sensitive biological activity.

4. Gas exchange experiment

A nutrient soup and CO_2 were added to a soil sample and followed for periods up to 12 days in the hope that microorganisms would generate the gases hydrogen, oxygen, nitrogen or methane, measured by chromatography. No gas exchanges of this type were measured.

Assembly of a Viking exobiology module.

A majority of experts concluded that the results of the Viking life-detection experiments were negative and largely due to chemical oxidation reactions. Nonetheless, the interpretation of the data is still the subject of active debate, because the nature of the Martian oxidizing compounds remains unexplained and without equivalents on Earth. This question will only be resolved through future missions. Either way, the Viking results have provided new bases for research on life on Mars. One of the principal objectives of the Phoenix mission, launched in 2007 (see page 200), was to look for organic compounds in the soil of the North Pole.

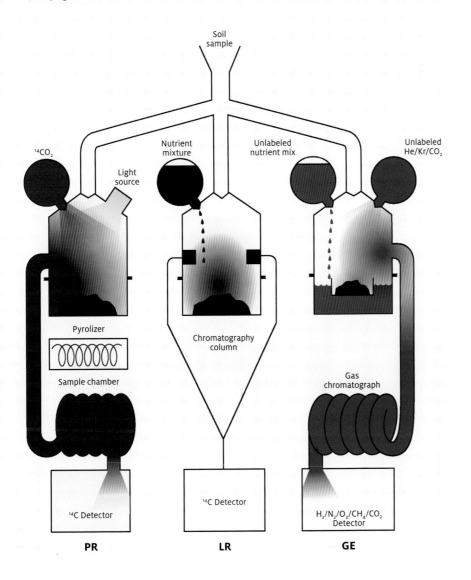

Soil sample

$^{14}CO_2$

Nutrient mixture

Unlabeled nutrient mix

Unlabeled He/Kr/CO$_2$

Light source

Pyrolizer

Sample chamber

Chromatography column

Gas chromatograph

^{14}C Detector

^{14}C Detector

H$_2$/N$_2$/O$_2$/CH$_4$/CO$_2$ Detector

PR

LR

GE

Diagram outlining the rationale of the experiment approach to detecting microbes in the Martian soil. PR: Pyrolytic Release, LR: Labeled Release, GE: Gas Exchange.

Martian geologic features
revealed by the Viking orbiters

The North Pole is covered by a cap composed of frozen carbon dioxide and water ice, whose thickness varies seasonally.

OPPOSITE PAGE, TOP
Composite image of Valles Marineris in its totality: 3,000 kilometers in length, with an average depth of 8 kilometers.

OPPOSITE PAGE, BOTTOM
Ice deposit on the South Pole.

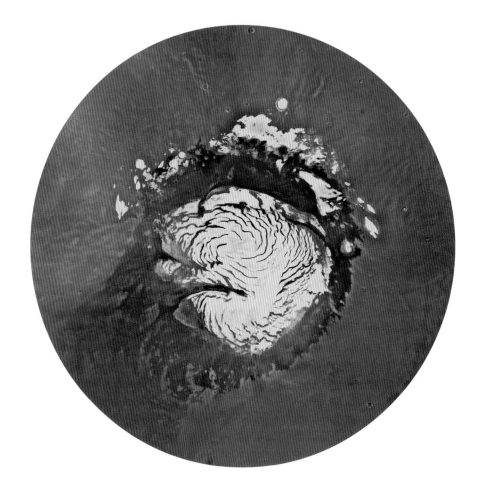

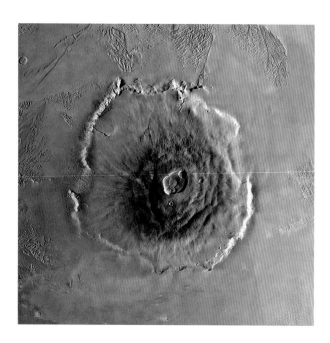

The gigantic Olympus Mons, the highest summit in the solar system, with an elevation of 22 kilometers and a circumference of 65 kilometers.

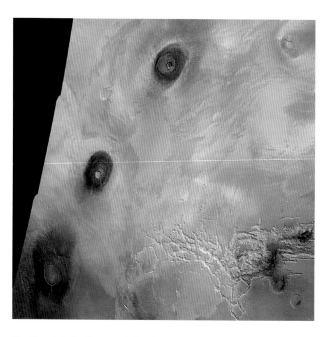

The Tharsis bulge forms a chain of enormous volcanoes: Ascraeus Mons (top, at north), Pavonis Mons and Arsia Mons.

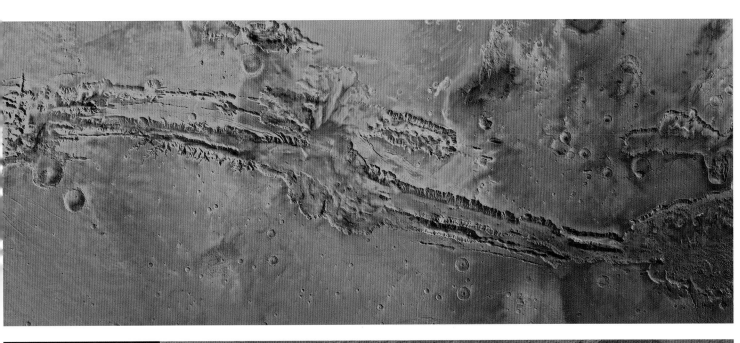

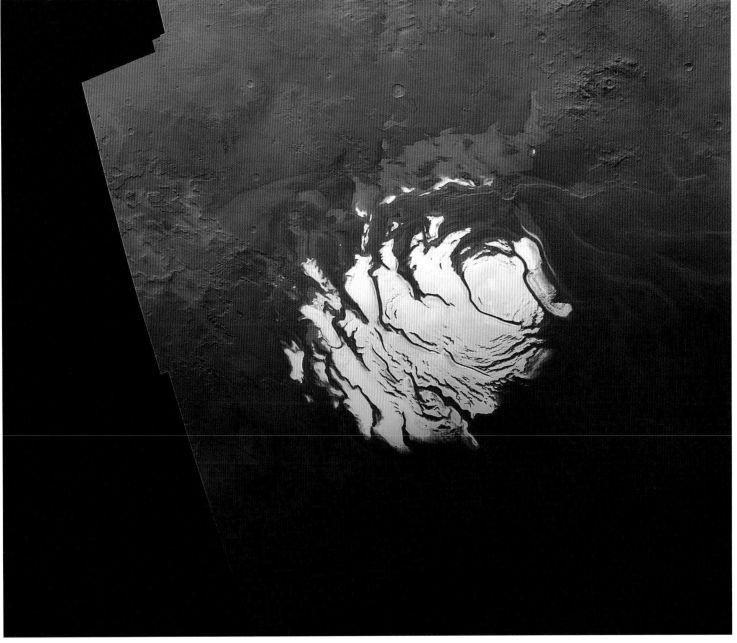

1988

Phobos 1 and 2

FIFTEEN YEARS AFTER their disastrous efforts of 1973, the Soviets undertook another mission to Mars with a new generation of space probes. This time the moon Phobos, an asteroid captured by Martian gravity, was the principal objective. The Mir space station had been launched in 1986, and the Soviets were training their cosmonauts for long-duration space trips. Phobos, in the not too distant future, could serve as a base for a crewed mission to Mars. The Phobos mission was designed to study the solar wind and cosmic rays, as well as the Martian atmosphere and surface. Two probes were intended to orbit the planet and undertake a low elevation (less than 100 kilometers) flyby of Phobos to

The Proton rocket with Phobos 1 on board is on track to the launch pad at the Baikonur Cosmodrome.

launch two science capsules. One was designed to plant itself in the ground (DAS in Russian, or LAL in English, for Long-term Automated Lander) and the other, a 43-kilogram module called Grasshopper, was to undertake several jumps on the surface. The probes were heavy (6,200 kg), with a scientific payload of 25 instruments. Their design had taken eight years, the fruit of a remarkable international effort between the Soviet Union during its "glasnost" period (the thawing of the cold war) and 14 other countries, including West Germany, Sweden and France. The United States also contributed in a major way by committing the NASA Deep Space Network to following the spacecraft on their voyages. Thanks to the powerful Proton-K rockets, Phobos 1 was launched on July 7, and Phobos 2 on July 12, 1988. Phobos 1 did not perform properly: it did not communicate with Earth as planned on September 2 and was lost.

Engineers learned that a computer command sent during flight on August 29 contained instructions to deactivate the attitude-control thrusters, which oriented the probe with respect to the Sun. These commands were part of a ground test sequence that, by mistake, had not been erased from the computer's memory before launch. One faulty line of code was enough to initiate a fatal series of events.

Phobos 2 inserted itself correctly into Martian orbit and started rounding its trajectory in order to approach Phobos. Following the first orbital correction on February 1, 1989, the spacecraft began to map the Martian surface in infrared light on February 11, 1989. It took a total of 38 high-resolution images of Mars and Phobos, but alas, on March 27, after approaching Phobos for the planned flyby at a 50-meter elevation, all radio contact with Earth was lost. The spacecraft was unable to reorient its solar panels correctly after the imaging session, thereby draining its reserve power supply and losing contact. It is likely that a burst of solar particles had damaged the on-board computer and affected its attitude-control commands.

Ultimately, this mission, which provided considerable data on the solar wind and spectacular images of Phobos, was only a partial success. The project to follow this one, Phobos-Grunt, was postponed several times, with an anticipated launch in 2011–2012. Its goal is to return a sample of Phobos to Earth. In the interim, on November 16, 1996, the Soviet probe Mars 96 was launched by a Proton-K rocket into a 166-kilometer-high circular parking orbit. The Block D-2 stage of the Proton was to accelerate the Russian spacecraft to a velocity of 3 kilometers per second, but malfunctioned, resulting in a 20 meters per second loss of speed. After the third orbit, Mars 96 re-entered the atmosphere and crashed into the Pacific, not far from the Easter Islands. Modeled after the Phobos probes, Mars 96 consisted of an orbiter, two penetrating landers and two rovers. The probe was an international cooperation involving NASA, ESA, France and Germany. After this catastrophe, the Russian astronautical community had no further opportunity to launch a new interplanetary probe until 2011. ♂

Model of the Phobos 2 probe.

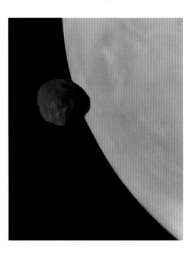

A faulty maneuver during its final approach to the Martian moon doomed the Phobos 2 mission.

BELOW
Infrared image of Mars taken by the panoramic camera on Phobos 2.

1996

Mars Global Surveyor

A FTER THE VIKING MISSION, great moments in space exploration followed, with the Voyager missions to the giant planets, the Mir space station, the Galileo probe to Jupiter, the launch of the Hubble Space Telescope and the European and Japanese missions to rendezvous with Halley's Comet. Nothing, however, was undertaken with respect to Mars. On September 25, 1992, the American probe Mars Observer was launched. Built along the lines of Earth-orbiting observation satellites and loaded with imaging and cartographic equipment, the exploratory potential of this mission was enormous. But, to the great dismay of the mission controllers, the spacecraft disappeared close to Mars, during its approach phase, after a voyage of 720 million kilometers. This fatal incident occurred after the command was sent to pressurize the fuel tank in preparation for ignition, to initiate orbital insertion. A review panel established that the most probable cause of failure was due to an explosion resulting from premature ignition and condensation of the oxidant in the fuel line. The propulsion system had been damaged by the extreme cold encountered during an interplanetary trip of several months.

After learning from the failure of Mars Observer, the United States returned to Mars in 1996 with the orbiting Mars Global Surveyor (MGS). Most of the instruments designed for Mars Observer were incorporated into MGS. Among other things, the scientific payload contained a digital camera for high-resolution imaging of the planet's surface, a magnetometer to study its magnetic field, a laser altimeter to measure cartographic relief and a thermal emission spectrometer to investigate soil and atmospheric properties.

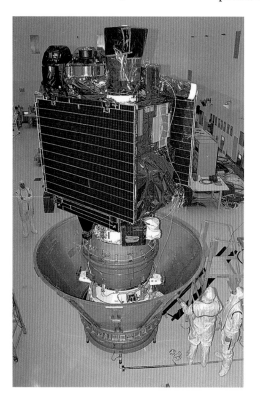

Preparation of the Mars Global Surveyor probe, showing the booster rocket to send it in the direction of its target.

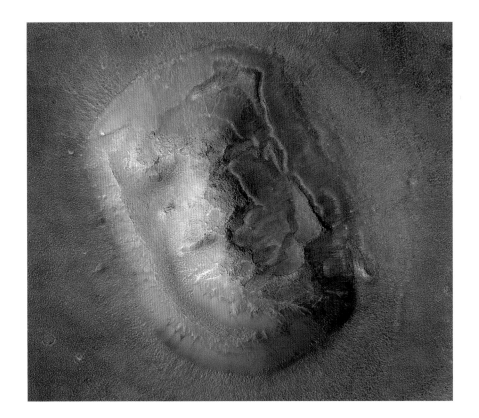

Mars Global Surveyor, shown in Martian orbit.

LEFT

People who loved Martian mythology were disappointed. Under very clear conditions, Mars Global Surveyor took this high-resolution image of the famous "face of Mars" in the Cydonia region. It turned out to be nothing more than a natural geologic formation.

MAIN COMPONENTS OF MARS GLOBAL SURVEYOR

1. High-gain antenna
2. Propellant tank.
3. Solar panel
4. Magnetometer
5. Star tracker
6. Thermal-emission spectrometer
7. Camera
8. Laser altimeter
9. Attitude-control thruster (4 in total)

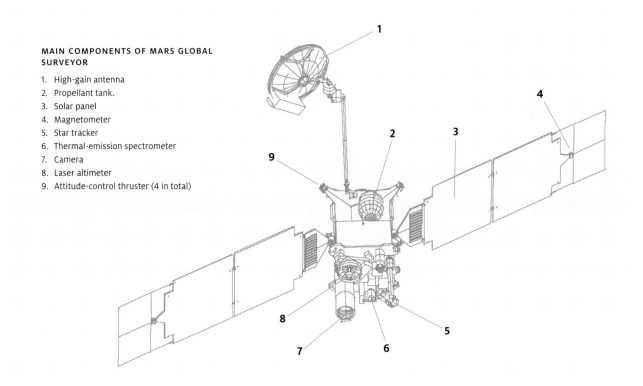

The MGS spacecraft weighed 975 kilograms: 595 kilograms for the bus and 380 kilograms for fuel. It was launched on November 7, 1996, by a Delta-7925 rocket from Cape Canaveral. It entered Mars orbit on September 11, 1997. The long journey to Mars (750 million kilometers over 300 days) proceeded without problems except for a solar panel that did not totally unfold during deployment. This would later cause difficulties for flight controllers during the probe's orbital insertion phase. As a result, an entirely novel method of aerobraking was used to save fuel for the mission. A critical factor was the pressure due to atmospheric friction that the solar panels were subjected to with each aerobraking pass. The process was consequently slowed down, thereby delaying the mission by more than two years. Surface mapping began on March 3, 1999, with the orbiter circling the planet every two hours at a mean altitude of 378 kilometers. The MGS mission was an outstanding success, making it possible to forget the failure of Mars Observer. More Mars data was collected than by all previous missions put together. The MOC camera (Mars Orbiter Camera) took more than 240,000 images of the surface. Using its laser altimeter, MGS was able to make an accurate global map of Mars, and for the first time it was possible to image channels and liquid outflows on the walls of some craters. Even more astonishing, images of the same areas taken over a number of years showed new outflows, thereby proving recent changes in the terrain. What are these outflow channels? On Earth, liquid water can cause similar erosion patterns, but on Mars, under much lower temperatures (around −50°C) and permanent low atmospheric pressure (less than 10 mbar), it is difficult to explain how water might pierce the surface. The debate continues.

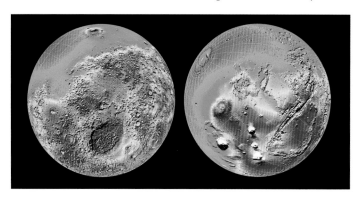

Topographic relief maps of Mars, produced by the laser altimeter aboard Mars Global Surveyor.

The last signal from MGS was transmitted on November 6, 2006, and it wasn't data, but a signal that the probe was going into hibernation. After that no further communications were established, for reasons that remain unclear to this day. What was clear was that this longest and most productive mission to Mars had to come to an end eventually. ♂

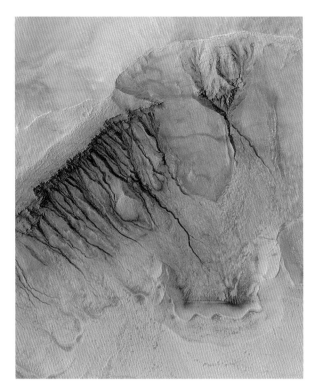

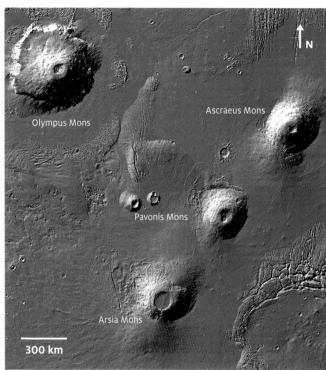

Mars Global Surveyor documented several outflow ravines within the walls of certain impact craters, this one in the Sirenum Terra region.

The four gigantic volcanoes on Mars in the Tharsis plateau.

BELOW

During its six years of operation, Mars Global Surveyor passed over the same areas several times, allowing it to observe recent outflows in the walls of this crater in the Centauri Montes region. Was this caused by liquid water?

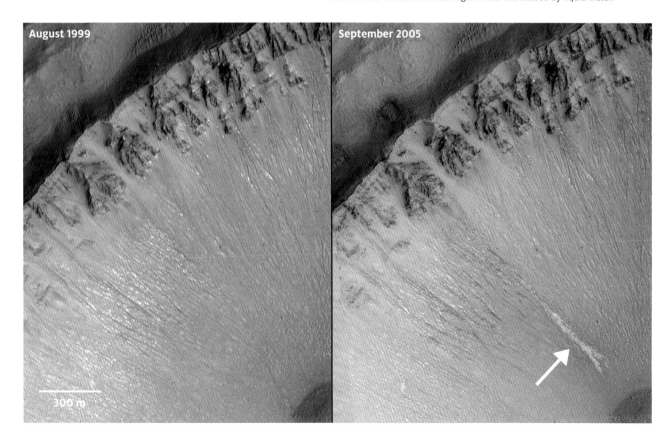

August 1999

September 2005

300 m

Mars Pathfinder

Planning for the Pathfinder mission, with the intention to land on Mars, began in 1992. Many were critical of the technical difficulties and higher risks inherent in NASA's credo in the 1990s: Faster, Better, Cheaper. It is true that there was no shortage of challenges for the team at the Jet Propulsion Laboratory that was responsible for the second project under the Discovery category (following project NEAR, page 328), a more technological than scientific mission. Consequently, the primary goal of the Pathfinder mission was to test several new technologies for planetary exploration: a high-speed, direct ballistic entry without first orbiting the planet; a safe landing on a rocky surface, thanks to inflatable airbags; and deployment of a semi-autonomous mobile rover. All that with a budget equivalent to one-fifth that of the Viking mission!

The Pathfinder mission was launched toward Mars atop a Delta II rocket on December 4, 1996, on a voyage lasting seven months. Aside from a problem with one of the solar trackers, which caused technicians much anxiety a few hours after launch but was quickly fixed, the trip between Earth and Mars went as planned, with a 27,000 kilometers per hour cruising speed. On July 4, 1997, the United States' Independence Day, Pathfinder began its descent to Mars.

BELOW, LEFT
A test of the inflatable airbags that served to cushion the landing of Mars Pathfinder.

BELOW, RIGHT
Before encapsulating Mars Pathfinder, Jet Propulsion Laboratory technicians refold Mars Pathfinder's petal-like solar panels. The rover Sojourner is solidly fixed onto one of the solar panels during the interplanetary trip.

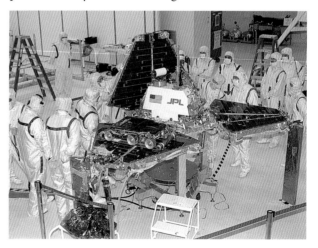

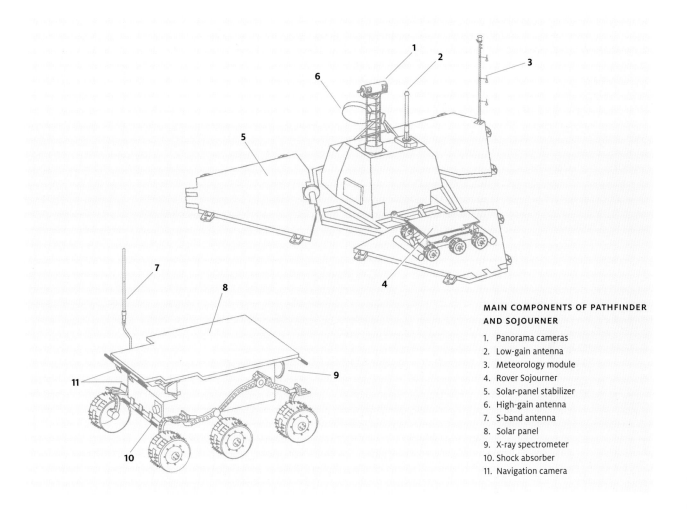

The little rover Sojourner, 63 centimeters long and weighing only 10.6 kilograms, was equipped with six independently steerable wheels.

Since radio signals traveling at the speed of light would take 20 minutes for the 233-million-kilometer round trip between Mars and Earth, the descent procedure had to be entirely automatic, with no possibility of human intervention. At 4 minutes and 30 seconds into the vertical descent, Pathfinder's velocity was reduced to 40 kilometers per hour, thanks to heat-shield braking and a supersonic parachute, followed by retrorocket stop at 20 meters above ground. The inflatable airbags were deployed eight seconds before impact, the drag chute released and the probe was in freefall within its protective cocoon. After a first bounce of more than 10 meters, followed by at least 15 more bounces, the package came to rest. The control team back on Earth was relieved when Pathfinder returned its first radio signal, which they interpreted as "I am alive!" After an absence of more than 20 years, NASA was back on Mars. Once the airbags were deflated and the lander "petals" opened to expose the solar panels, Pathfinder began the mission on the second day, by transmitting an image of the landing site and deploying the rover.

The lander's equipment consisted primarily of a widefield stereoscopic camera and a meteorological station, while the little six-wheeled rover (weighing only 10.6 kilograms) housed navigation cameras and an X-ray spectrometer mounted on an arm, used for chemical analysis of Martian rocks. The selection of the landing site, the alluvial valley Ares Vallis, in the Chryse Planitia region of the northern hemisphere, was based on its varied geology. Images obtained by the Viking orbiter showed an extended rocky surface — probably the result of extensive flooding that carried sedimentary and volcanic rocks over large distances.

The first Pathfinder image, showing that the lander petals had opened and Sojourner was in good condition after landing.

OPPOSITE PAGE

In the course of its mission in the Ares Vallis region, the rover Sojourner roamed 100 meters and analyzed several magma rocks, similar in composition to terrestrial granite.

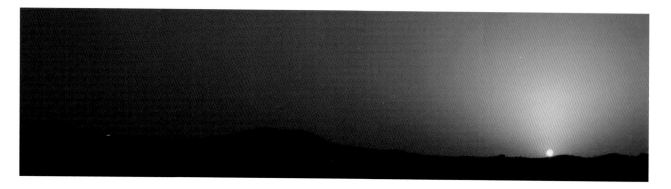

Martian sunset, photographed on the 24th day of the mission.

BELOW

A 360° panoramic view of the Pathfinder landing site. The Twin Peaks mountains, visible on the horizon, are about 2 kilometers distant.

The Pathfinder lander was subsequently named the Carl Sagan Memorial Station, in honor of the celebrated American astronomer who passed away in 1996. The rover was named Sojourner in honor of Sojourner Truth (1797–1883), a militant American abolitionist who did much to advance women's rights. On the third day, Sojourner descended its access ramp, marking the first time in history that a rover moved about the surface of another planet. Carl Sagan had dreamt of such a moment 21 years before: "The problem is that machines like Viking are immobile … We therefore desperately need a mobile robot." The lander relayed information between the Earth and Sojourner, which moved at a very modest rate of 1 centimeter per second. The Pathfinder mission is considered a great success. More than 16,500 images were taken by the lander and 550 by Sojourner, which also analyzed the elemental composition of 15 samples of Martian rocks.

For the first time, thanks to the Internet, many people on Earth were able to follow Sojourner's adventures on Mars almost in real time, despite its distance of more than 100 million kilometers. This helped make the rover an international mascot. There were about 566 million Internet visits during the first month of the mission: 47 million on July 8 alone. Scientific results obtained through the X-ray spectrometer and imaging confirmed that rocks at the Ares Vallis site consist of andesite and basalt, like terrestrial volcanic rocks. Unfortunately, the ubiquitous Martian dust affected the geologic analyses by masking details and contaminating surfaces, thereby limiting the quality of the observations.

Designed to function for about a week, the Pathfinder mission ultimately lasted three months before communications with Earth ceased for unknown reasons. In case of difficulties, Sojourner was programmed to circle its home base and await instructions. It's likely that the orphaned robot tried desperately to capture a signal for several days before succumbing to the implacable Martian cold. On a positive note, Pathfinder's innovative method of entry and descent to the Martian surface, as well as the concept of a rover, were used on a much larger scale a few years later for the Mars Exploration Rovers. ♂

2001

Mars Odyssey

NASA utilized the same design and most components of the ill-fated Mars Climate Orbiter in the construction of Mars Odyssey.

OPPOSITE PAGE
Artist's rendition of Mars Odyssey in orbit.

THE MARS SURVEYOR 98 MISSION, with its sophisticated Mars Climate Orbiter and Mars Polar Lander, was a double disaster for NASA. On September 23, 1999, the Mars Climate Orbiter probe, launched December 11, 1998, approached the orbital insertion phase only to burn up in the upper atmosphere of Mars. This loss was due to a mismatch between imperial and metric units of measurement used by different teams in calculating the spacecraft's altitude. Just a few months later, on December 3, 1999, the Mars Polar Lander, launched on January 3, 1999, also crashed into the red planet. Humiliated, embarrassed and criticized by the media, NASA halted all future Mars missions and revised its plans. Failure was no longer an option. Great care was taken in planning the mission of the orbiting spacecraft Mars Odyssey, so named with a nod to the celebrated science fiction movie, *2001: A Space Odyssey*, directed by Stanley Kubrick and based on the work of Sir Arthur C. Clarke. For this mission, NASA decided to re-use the basic elements of Mars Climate Orbiter, which failed due to navigational problems but was otherwise solidly designed. Mars Odyssey would not have a lander, however, since that bore too many similarities to the defunct Mars Polar lander. Mars Odyssey was launched atop a Delta II-7925 rocket on April 7, 2001, from Cape Canaveral, and entered Mars orbit on October 24, after a journey of more than six months and 460 million kilometers. The spacecraft would spend an additional 3 long months and 322 aerobraking tours of the planet before attaining its final orbit. However, this excruciatingly slow approach, first utilized successfully by Mars Global Surveyor, helped save more than 200 kilograms of propellant fuel! The primary mission began on February 19, 2002, and remains operational to this day. Mars Odyssey had two main objectives: map regions on Mars rich in hydrogen (and hence water), and

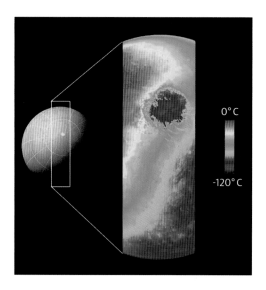

serve as a communications relay to Earth for future landers. Mineral mapping is made possible by two types of spectrometers on the spacecraft: a high-resolution Thermal Emission Imaging System (THEMIS), operating at longer visible and infrared wavelengths, and a gamma-ray detector. Two additional instruments round out the array, a neutron spectrometer and MARIE (MArs Radiation EnvIronment Experiment), designed to evaluate the radiation risks future astronauts might be exposed to. The THEMIS imager has provided many thousands of images of the planet, covering almost the entire surface to a resolution of 20 meters. The presence in the equatorial plains of extensive areas of hematite, which is usually formed in the presence of water on Earth, persuaded scientists to send the Mars rovers Spirit and Opportunity (see page 176) to investigate that geology directly. Most of the images taken by the rovers, as well as Phoenix, were first sent to Mars Odyssey and then relayed to Earth. Since Mars Odyssey, the decision has been made to place relay orbiters around Mars to help out with future missions to the planet. ♂

Surface temperatures at the North Pole of Mars, as measured by the THEMIS spectro-imager.

BELOW

Map of Mars showing potential locations for water. The locations of various landers are also shown:
G : Mars Rover Spirit at Gusev
M : Mars Rover Opportunity at Meridiani
PF : Pathfinder
V1 : Viking 1
V2 : Viking 2.

H$_2$O (%)

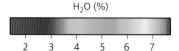

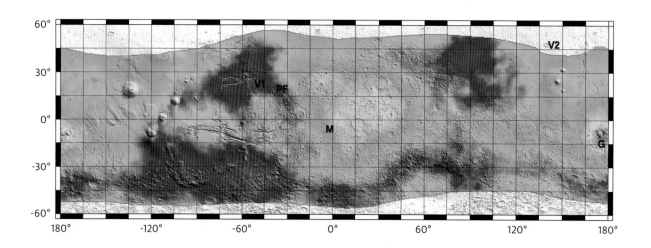

ARTHUR C. CLARKE

(1917–2008)

Sir Arthur Charles Clarke was a British inventor and science fiction author. During the Second World War, he contributed to the development of an early warning radar defence system, an important factor in the success of the Battle of Britain. He dedicated himself to writing in 1951. He became famous in 1968 with the movie *2001: A Space Odyssey*. Directed by Stanley Kubrick, it was based on the novel by the same title, an adaptation of the short story "The Sentinel." A prolific writer, Clarke also conceived the idea of a satellite in geosynchronous orbit, now termed the Clarke orbit, which he proposed in the October 1945 issue of the magazine *Wireless World*. He retired to Sri Lanka in 1956, where he founded a center for deep-sea diving, which, unfortunately, was destroyed by the tsunami of December 26, 2004.

Two possibilities exist: either we are alone in the Universe or we are not. Both are equally terrifying.

Clarke's three "Laws"

1. When a distinguished but elderly scientist states that something is possible, he is almost certainly right; when he states that something is impossible, he is probably wrong.

2. The only way of discovering the limits of the possible is to venture a little way past them into the impossible.

3. Any sufficiently advanced technology is indistinguishable from magic.

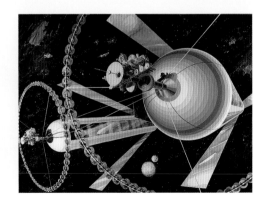

Mars Exploration Rovers

I T IS OBVIOUS that mobility is all-important when it comes to exploring the Martian environment, and Pathfinder was the proof that it is now possible to remotely control rovers on the surface of Mars. In July 2000 NASA announced approval for the Jet Propulsion Lab plans to send rovers to investigate the geology of Mars, looking for traces of water, with a launch in 2003. A concurrent proposal, by Lockheed Martin Astronautics, to send an orbiter with a very powerful camera, though less risky, but also less exciting, would become reality a few years later. That would become the Mars Reconnaissance Orbiter, launched in 2005 (see page 192). With the Mars Exploration Rovers, NASA returned to its longstanding practice of sending several identical interplanetary spacecraft, like Mariner 6 and 7, Pioneer 10 and 11, Voyager 1 and 2, and Viking 1 and 2. Missions like these helped reduce the costs of development and maximize scientific returns, while also lowering risks. The two rovers, about the size of a 180-kilogram golf cart, were officially named Spirit (MER-A) and Opportunity (MER-B). These names were provided through a contest won by Sofi Collis, a nine-year-old Russian-American orphan, who wrote a poignant letter ending with: "In America, I can make all my dreams come true ... Thank you for the spirit and opportunity." Spirit was launched on June 10, 2003, and Opportunity on July 7, 2003. Spirit was propelled by a standard Delta II 7925 rocket, while Opportunity, launched later in the launch window, needed a version H (Heavy) lifter rocket. Thanks to the exceptionally close approach between the Earth and Mars that year, the two probes had to travel "only" 56 million kilometers and the trip required "only" six months. Spirit landed on January 3, 2004, in Gusev Crater, which resembles a dry lake bed, while Opportunity landed January 24 on the other side of the planet, at Meridiani Planum, a hematite-rich plain. The landing sites were chosen based on Mars Global Surveyor observations.

OPPOSITE PAGE, TOP

Opportunity's protective petals are opened for final adjustments. Its solar panels are shown in the closed configuration.

OPPOSITE PAGE, BOTTOM

The rover in its thermal shield is placed on a turntable for technical inspections.

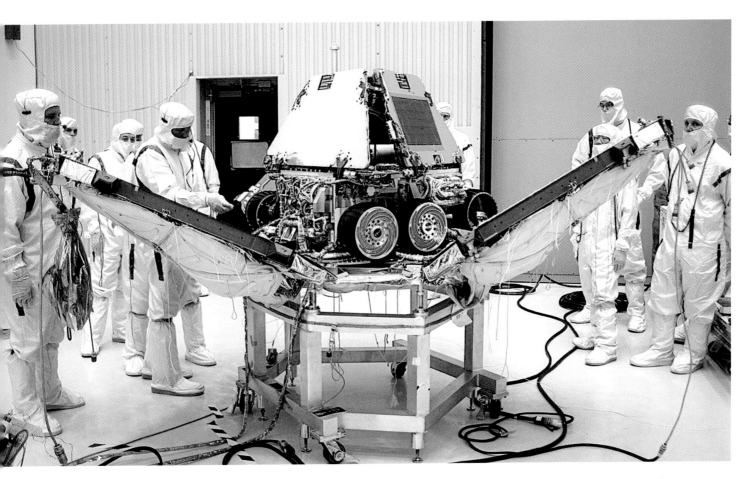

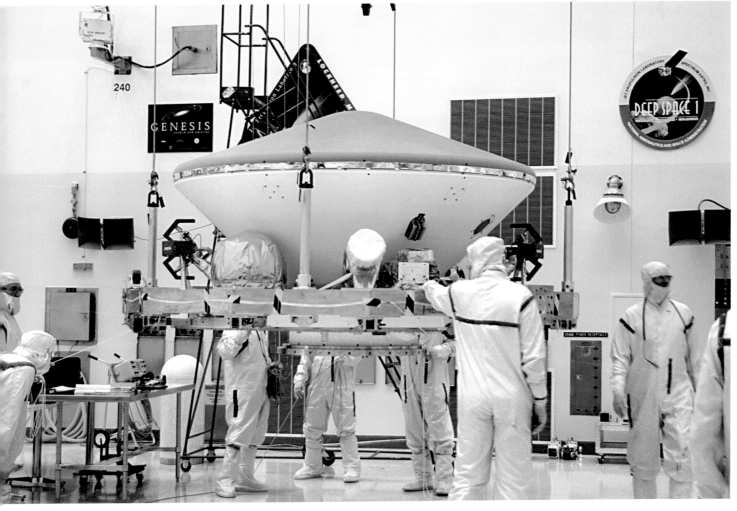

Wind-tunnel test of the supersonic parachute.

OPPOSITE PAGE
Artist's rendition of the rover Opportunity on the surface of Mars.

Equipped with three types of radio antennas (a high-gain parabolic, an omnidirectional low-gain and an omnidirectional UHF), communications between Earth and the rovers were limited to one a day. That meant that each rover had to be able to move in autonomous mode, without direct control. To accomplish that, each robot had to be programmed with artificial intelligence capable of analyzing surface images taken by its navigation camera, to avoid obstacles and define the best path to follow. If it encountered difficulties in making such a decision, the rovers were programmed to go into stop mode and await instructions from Earth. Like their precursor Sojourner, the rovers have six independently movable wheels with a rocker-bogie type suspension. Their solar panels are capable of providing up to 140 watts to recharge batteries. This limited their movement to daytime only, since they are totally dependent on solar energy. Because of this, rover activities have to be carefully managed. Solar illumination angles vary seasonally and in relation to the amount of dust suspended in the Martian atmosphere, as well as the amount of dust accumulation on the solar panels, which all impact the amount of energy available.

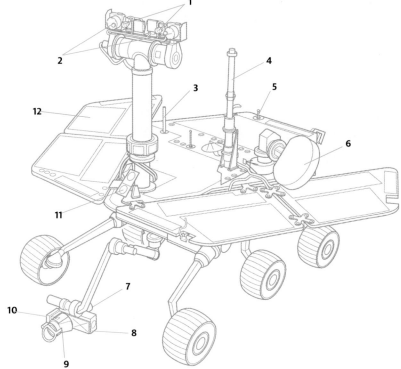

PRINCIPAL COMPONENTS OF THE ROVER SPIRIT

1. Navigation cameras
2. Panorama camera
3. UHF antenna
4. Low-gain antenna
5. Calibration target
6. High-gain antenna
7. APXR spectrometer
8. Mössbauer spectrometer
9. Abrasion wheel
10. Microscope
11. Magnets
12. Solar panels

The two Mars rovers were programmed for all phases of their mission. Unlike Sojourner, whose relay base was also equipped with a camera and meteorological station, Spirit and Opportunity left their landers and protective cocoons behind, and were capable of traveling several dozen meters per day in search of evidence that water was abundant on Mars at some time in its past. To this end, each rover carried a panoply of scientific equipment and tools, including Pancam, a digital camera for high-resolution 3-D panoramic imaging, and a thermal emission spectro-imager to assess the composition of the surrounding terrain. It was also outfitted with an articulating arm equipped with Athena, a palette of robotic instruments that the rover can place in contact with rock or soil samples. Athena contained a digital microscope, an X-ray spectrometer (APX, for Alpha Proton X-Ray Spectrometer) similar to the one on Sojourner, a Mössbauer spectrometer to study the elemental composition of the target samples and a diamond Rock Abrasion Tool (RAT) to grind into rock specimens and remove surface dust. To ensure the validity of their data, all instruments could be independently calibrated. The calibration target of the Pancam was particularly important in this regard and served as a sundial. Designed to adjust for luminosity and color calibration for all images, it became the most photographed object on Mars!

Panoramic view of Victoria Crater taken by Opportunity from the Cape Verde promontory. The furthest wall lies about 800 meters from the rover (see page 194–195). This series of images was taken between October 16 (the 970th Martian day of the mission) and November 7, 2006.

OPPOSITE PAGE
The lander, with its protective cocoon of inflatable balloons, functioned perfectly during the mission.

Here is another bit of evidence that there was abundant water on Mars in the past: sulfur-rich rocks, probably submerged, exhibit stratification patterns, wherein concretion spherules formed.

TOP RIGHT
An articulating robotic arm, shown touching a sample, made the rovers into sophisticated geologists.

RIGHT
The "blueberry" spherules discovered by Opportunity at its landing site contain hematite, an iron oxide that forms mainly in the presence of water on Earth.

The scientific results accumulated by the two rovers confirm that water played an important role in past Martian geology (areology). However, the presence of hematite in the soil, though evidence in favor of the presence of water, is not conclusive proof. This iron-oxide rich mineral is so abundant on Mars at present that it gives the planet its characteristic rusty color. On March 2, 2004, NASA announced officially that Opportunity's instruments had provided convincing evidence that the rocks found at the rover's landing site were submerged in the past, because they contained jarosite, an iron sulfite mineral. What happened to Martian water continues to be unexplained at present, but this mission supports the hypothesis that water flooded large portions of Meridiani Planum and played an important role in its geology several billion years ago.

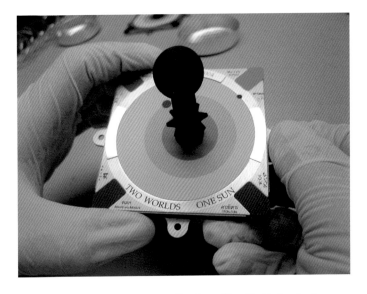

The slogan "Two Worlds One Sun" decorates the calibration target of the Pancam. The target also serves as a sundial and contains the name Mars in 22 languages.

While on its way to the crater Erebus, on April 26, 2005, Opportunity became trapped when all its six wheels got stuck up to 80 percent in very fine sand. It took JPL engineers five weeks to gradually free it, centimeter by centimeter. After that it began roving again and remains active to this day. As of December 2010, Spirit and Opportunity had logged 7.7 and 26.8 kilometers, respectively. This is particularly impressive, given that their initial life expectancies were about 90 days and their range of operation about 600 meters! Irretrievably stuck, Spirit has become a fixed station that has not communicated with Earth since March 2010. At the time of writing, Opportunity is continuing on its way toward the crater Endeavour. After several years of operation in an environment as hostile as Mars, the rovers are a testament to the remarkable work accomplished by the engineers and scientists at Jet Propulsion Laboratory and NASA. ♂

2003

Mars Express

BELOW, LEFT

Mars Express, launched in 2003, is the Eurpoean Space Agency's first planetary mission.

BELOW, RIGHT

The imposing Soyuz rocket is moved to the launch pad at Baikonur to launch Mars Express, on June 2, 2003.

OPPOSITE PAGE, TOP

Mars Express in its approach phase. The Beagle 2 lander is still attached to the orbiter.

OPPOSITE PAGE, BOTTOM

The Beagle 2 lander, as it might have looked on the surface of Mars.

MARS EXPRESS is the European Space Agency's first mission to explore the red planet. The name "Express" underscores both the speed with which the project was conceived and implemented, and the relatively short time it took to traverse the Earth–Mars distance in 2003. The project saved considerable time by incorporating several instruments used by the failed Soviet Mars 96 mission and by utilizing the overall framework of the European Rosetta spacecraft (see page 313). In order to minimize costs, this same framework was chosen for the Venus Express probe (see page 124).

Mars Express was launched from Baikonur on June 2, 2003, on board a Soviet Soyuz-Fregat rocket. The reliability of the Soyuz rocket (originally the R-7) is simply remarkable, with 1,500 successful launches since 1963 and a nearly 98 percent success rate. After reaching a 200-kilometer Earth parking orbit, the Mars Express probe was launched toward its target by the multiple-firing Fregat booster stage. The mission goal was to orbit Mars and send the lander, Beagle 2, to the surface.

The name Beagle 2 comes from the ship *HMS Beagle*, on which British biologist Charles Darwin circumnavigated the world in the 19th century. The principal goals of the mission were to study the atmosphere of Mars, its surface and subsurface, from orbit. The lander, Beagle 2, was to complete the mission by looking for life (exobiology research) in the Martian soil.

The six-month long and 400-million-kilometer Earth to Mars transition was not routine. Technicians at ESOC (the ESA Operations Center) in Darmstadt, Germany, were plagued by star navigation problems, a bad cable connection that reduced solar panel capabilities, and a gigantic solar eruption that hit the spacecraft with full force. After releasing Beagle 2 toward Mars, Mars Express returned to its original trajectory and began orbital insertion by firing its retrorockets to slow the spacecraft to 5.2 kilometers per second, the minimal escape velocity from Martian gravity. On December 25, Mars Express entered a highly elliptical orbit around Mars, reaching a final orbit of 298 by 11,560 kilometers. Highly elliptical orbits are not as good for mapping purposes as more circular orbits. Since the angle of view varies greatly as a function of orbital position, this makes image comparisons difficult. There was not enough fuel on board to adjust the orbit to a more circular one, and the European technicians had not yet mastered the delicate technique of aerobraking. As a result, each orbit consisted of an observational stage at low elevation above the planet, followed by an X-band radio communication stage that required a rotation of the probe toward Earth. In addition to the two large ESA radio antennas in New Norcia, Australia and in Madrid, the NASA Deep Space Network also assisted with communications. Among the primary Mars Express instruments were a high-resolution stereoscopic camera to map the planet; a visible and infrared light spectrometer to look for water and study the mineral composition of the surface;

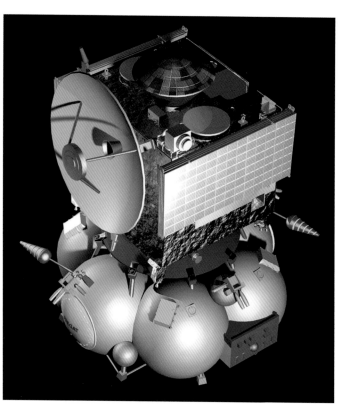

The Mars Express probe in launch configuration, atop its Fregat booster rocket.

an ultraviolet-infrared spectrometer to study the composition of the atmosphere; and a 20-meter radar antenna to penetrate below ground and look for ice deposits as far down as 5 kilometers. Images obtained by the stereoscopic camera were spectacular and the radar detected permafrost deposits under the poles. The spectrometer detected traces of methane and ammonia in the atmosphere. This is intriguing, because both of these light gasses would be expected to escape quickly from Mars. Their presence might suggest active production, through biological or volcanic processes. In addition to its own objectives, Mars Express was designed to serve as a radio relay in support of Mars landers, like Beagle 2 and Phoenix, thanks to the adoption of the standard international space-communications system Proximity-1. Mars Express was a total success from both a scientific and an astronautical perspective. The experience gained in the course of this first interplanetary mission will serve ESA well with future space exploration projects.

The Beagle 2 lander, developed by a British team led by Professor Colin Pillinger and funded by public and private support, was less fortunate. It was a last-minute addition to the Mars Express mission, in an effort to optimize science returns. Mars Express was placed on a collision path with Mars on December 19, 2003, to release the Beagle 2 module, which entered the Martian atmosphere in ballistic fashion at 20,000 kilometers per hour a few days later. The lander's entry and descent system was similar to that of Pathfinder, with a heat shield, parachute and inflatable balloons to cushion the final landing. Unfortunately, while its release from Mars Express went as planned, according to the video images transmitted, Beagle 2 gave no further sign of life and probably crashed somewhere in the Isidis Planitia area on December 24. Its loss was officially announced on February 6, 2004. Despite attempts by both the Mars Global Surveyor and Mars Reconnaissance Obiter teams to locate it, the exact crash location of Beagle 2 remains undetermined. More serious still,

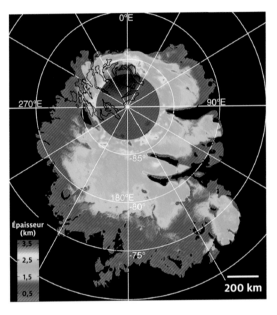

MARSIS (Mars Advanced Radar for Subsurface and Ionosphere Sounding) radar map of the South Pole of Mars, showing the thickness of the ice layer. It was calculated that if this mass of ice were to melt, the planet would be covered by 12 meters of water.

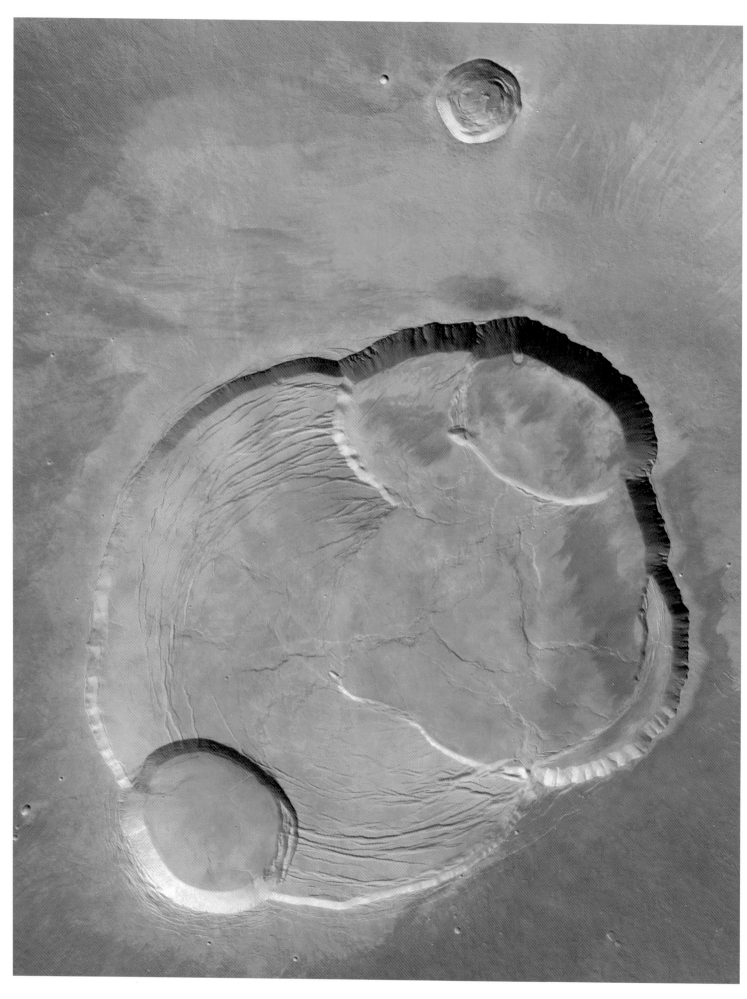

the exact cause of failure has not been established with certainty. When the rovers Spirit and Opportunity landed, a few days after Beagle 2, they reported an abnormal drop in atmospheric pressure. An unexpected, large drop in pressure might have affected the braking capacity of Beagle 2. Unlike Pathfinder and the two rovers, Beagle 2 did not have retrorockets coupled to an altimeter to actively slow it down after the release of its parachute. Although there are rumors of Beagle 3, clearly additional precautions will have to be taken to avoid another landing mishap, although that will always be a risky process. From an engineering standpoint, due to strict weight limitation (a maximum of 60 kilograms), the design team had to resort to great ingenuity to make provisions for the scientific instrument payload. Beagle 2 resembled a flat saucer. After landing horizontally on the surface, it was designed to open up and deploy four solar panels and a UHF antenna, and then to activate a movable robotic arm with a multitool PAW (Position Adjustable Workbench).

OPPOSITE PAGE
Details of the caldera of the Olympus Mons volcano.

BELOW
Angular view of an ice field inside the crater Vastitas Borealis.

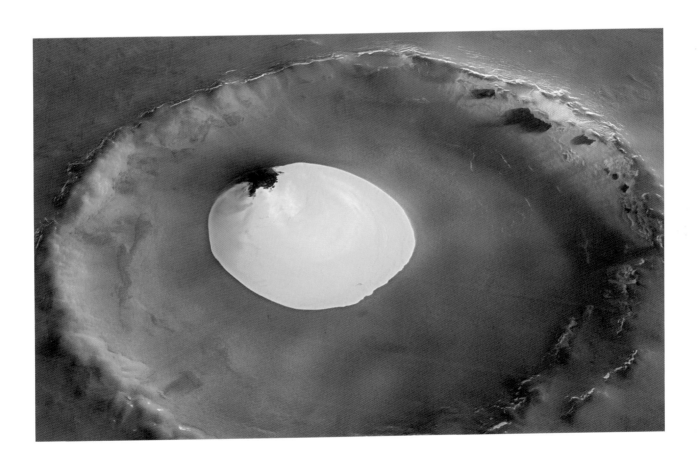

On July 23, 2008, Mars Express transmitted this very detailed image of the moon Phobos, taken during a flyby from less than 100 kilometers at a velocity of 3 kilometers per second.

OPPOSITE PAGE

The high-resolution stereoscopic images taken by Mars Express provided outstanding topographic results. Shown here are the views of Rupes Tenuis at the North Pole (top) and Noctis Labyrinthus (bottom). Compare this to the image on page 143, obtained by Mariner 9 in 1972.

The multitool was equipped with a stereoscopic camera, a microscope, a sample collector, a drill, an X-ray spectrometer and a Mössbauer spectrometer to detect iron-rich rocks. One of Beagle's instruments to look for evidence of life on Mars was GAP (Gas Analysis Package), based on a mass spectrometer. Basically, GAP was designed to analyze the chemical composition of the atmosphere and samples of rocks and soil for evidence of molecules of biological origin. It is likely that the Martian surface is totally sterile, due to a combination of chemical oxidation and the constant exposure to solar ultraviolet radiation. That is why Beagle 2 was also equipped with a digging tool, PLUTO (PLanetary Undersurface TOol), to collect and analyze samples below ground. After the failure of Beagle 2, the search for life on Mars remains a high priority for future missions. ♂

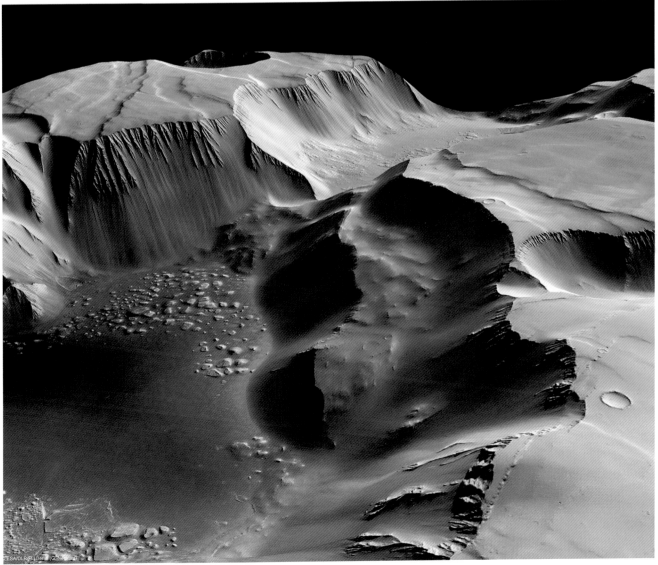

ESA/DLR/FU Berlin (G. Neukum)

Mars Reconnaissance Orbiter

Inspection of Mars Reconnaissance Orbiter's solar panels during an electromagnetic interference test.

OPPOSITE PAGE

Mars Reconnaissance Orbiter, shown at low elevation above the Martian North Pole.

BELOW

Against the background of the Atlantic Ocean, MRO is successfully launched by an Atlas V rocket on August 12, 2005.

NASA'S MARS RECONNAISSANCE ORBITER (MRO) is a multifaceted satellite designed to increase our ability to observe and explore Mars for future human missions. Weighing 2.18 metric tons and standing 7 meters high, with a fully extended solar panel wingspan of 13.4 meters, Mars Reconnaissance Orbiter is by far the largest space probe sent to the red planet. MRO was launched by an Atlas V-410 rocket from Cape Canaveral on August 25, 2005. When it arrived at Mars in 2006, it joined five other probes already in operation: Mars Express, Mars Odyssey, Mars Global Surveyor and the two Mars Exploration Rovers. After a seven-month aerobraking procedure, MRO descended from a highly elliptical orbit (426 by 44,500 kilometers) into a two-hour polar orbit between 250 and 316 kilometers in altitude. One of the most exciting objectives of the mission was to map the Martian surface to resolve objects less than one meter in size. Such unparalleled precision was possible thanks to a 50 cm aperture telescope and the HiRISE (High Resolution Imaging Science Equipment) digital imager, a first of its kind in spacecraft technology. A single, 800-megapixel (20,000 x 40,000 pixels) color image captured by HiRISE fills 16.4 gigabytes of memory before compression to 5 gigabytes for transmission to Earth. In addition to providing a wealth of geological data, these highly detailed images will help locate the best landing sites for future explorations, both in terms of safety and scientific interest. In this regard, the HiRISE camera can take paired stereoscopic images to better evaluate the terrain of potential sites. Another camera, with 350 mm focal length, was included for wide-angle views of areas imaged by HiRISE. A third camera, MARCI (Mars Color Imager),

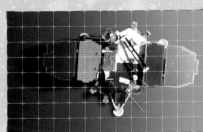

MARS POLAR LANDER

Resolution of Mars Global Surveyor camera

Resolution of the Mars Reconnaissance Orbiter camera

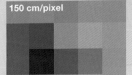

150 cm/pixel

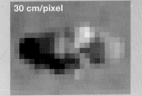

30 cm/pixel

The HiRISE camera, with its 50 cm aperture telescope mirror, is the key instrument of the Mars Reconnaissance Orbiter mission.

From altitudes between 200 and 400 kilometers, HiRISE can discern objects as small as Martian landers.

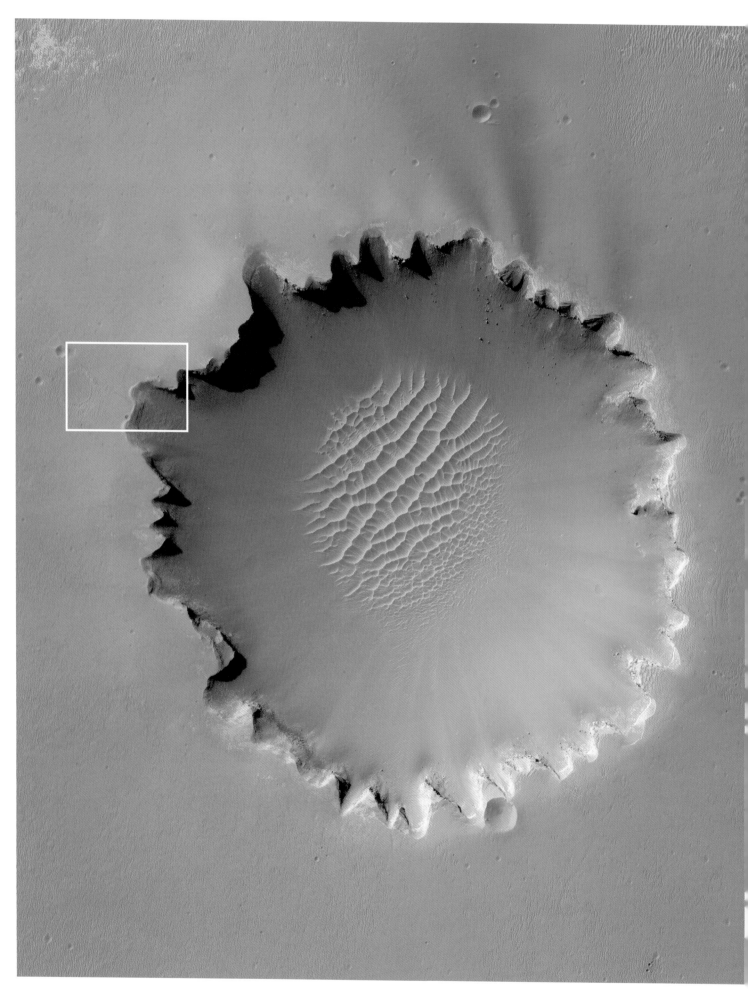

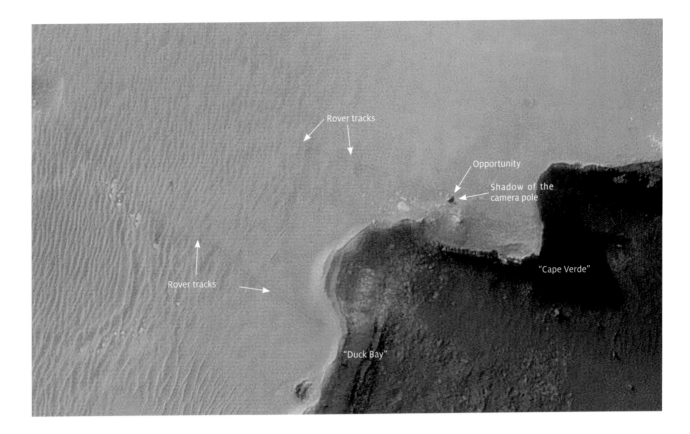

Rover tracks

Opportunity

Shadow of the camera pole

"Cape Verde"

Rover tracks

"Duck Bay"

takes 84 images daily of Mars to generate global meteorological maps. Such maps have made it possible to better understand seasonal and annual atmospheric changes on Mars, in anticipation of future missions, whether robotic or crewed. MRO is also equipped with several types of spectrometers, in order to better characterize the mineralogy, temperature and humidity of the Martian surface. In addition, the Italian Space Agency provided SHARAD (Shallow Subsurface Radar), which will look for evidence of sub-ground ice deposits to a depth of 1 kilometer, to verify the results from Mars Express. While SHARAD provides greater resolution (10 meters) than MARIS (50 to 100 meters), the latter can penetrate the Martian soil much deeper, to 5 kilometers. In addition, MRO is equipped with novel equipment for UHF communication with other spacecraft during approach, the landing phase or for surface operation on Mars. MRO relays information to the two rovers, Spirit and Opportunity, as well as the Phoenix probe. With its 3-meter diameter antenna, the MRO telecommunications system is the most advanced and powerful ever housed on an interplanetary spacecraft. To demonstrate the power of this technology, a Ka band transmission at a frequency of 32 GHz can theoretically relay

OPPOSITE PAGE AND ABOVE

The high-quality images captured by HiRISE make it possible to distinguish not only the rover Opportunity, near Victoria Crater, an 800-meter-diameter impact feature, but also the shadow of its camera pole and wheel tracks in the soil!

DS1
(Comets)
15
Gigabytes

Mars
Odyssey
1012
Gigabytes

Mars Global
Surveyor
1759
Gigabytes

Cassini
(Saturn)
2550
Gigabytes

Magellan
(Venus)
3740
Gigabytes

MRO
43
Terabytes

34
Terabytes
needed by
NASA

A deluge of data: the Mars Reconnaissance Orbiter will transmit three times more data than five other interplanetary missions combined.

OPPOSITE PAGE, TOP

Sediments of dust and sand have sculpted and recoded one of the valleys of Ius Chasma in the Valles Marineris area.

OPPOSITE PAGE, BOTTOM LEFT

Sublimation (the transformation of ice into gas) of carbon dioxide during the polar spring has created these bushy-looking features, which have no terrestrial counterpart.

OPPOSITE PAGE, BOTTOM RIGHT

These wave-like rock erosions in Becquerel Crater reflect the cyclical variations in the planet's rotational axis.

6 megabytes per second, 10 times higher than any previous orbiter. All complex functions are controlled by an on-board computer with a 32-bit RAD750, 133 MHz processor, an upgraded version of the PowerPC 750 or G3 processor, known to Macintosh and VxWorks users. Although this processor may seem obsolete in comparison to current computers, reliability and resilience are the most important factors in space. The system's flash memory of 160 gigabytes is sufficient (though not great) in terms of the size of the images it must deal with. Thanks to its imaging equipment and great telecommunications capabilities, MRO has already provided more results than all previous Mars missions combined! In summary, MRO is an outstanding spacecraft with exceptional scientific potential. Everyone is hoping that the mission will be extended, at least to the stage where it can recover traces of the ill-fated Mars Polar lander and Beagle 2. ♂

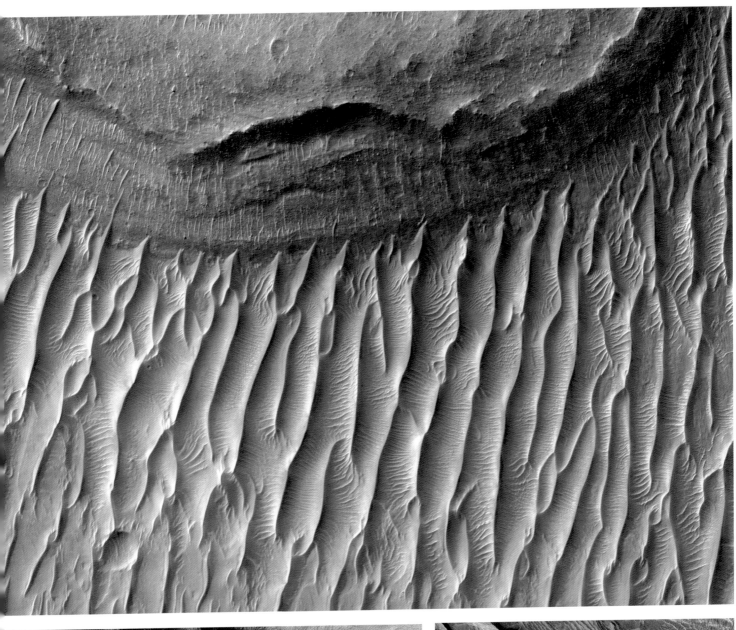

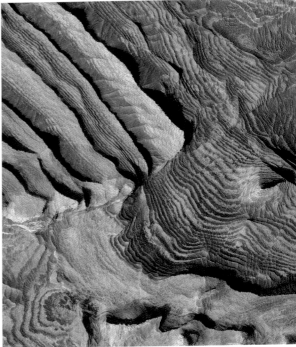

The large crater Stickney on Phobos is 9 kilometers in diameter. The lines and craters on Phobos are the result of secondary impacts caused by ejecta and meteoric collisions on Mars.

OPPOSITE PAGE

Erosion of the upper layers has revealed ice-rich materials within Chasma Boreale at the Martian North Pole.

Phoenix

A technician inspects the articulated robotic arm on the Phoenix lander.

Tʜᴇ ᴘʜᴏᴇɴɪx is a mythical bird that rose from its ashes. This appropriately named spacecraft rose from the ashes of the cancelled 2001 mission, Mars Surveyor, after the disastrous loss of Mars Polar Lander in 2000. The Mars Surveyor 2001 lander had been carefully mothballed by its manufacturer, Lockheed Martin, for all these years, which made it possible to use many of its structural components, thereby lowering Phoenix costs. The Phoenix mission was the first of NASA's Scout program, intended to send smaller and less expensive space probes to Mars. This was made possible through an innovative collaboration among the University of Arizona for scientific instruments; Jet Propulsion Laboratory (JPL) and NASA to run the mission; Lockheed Martin Space Systems for production of the spacecraft; as well as

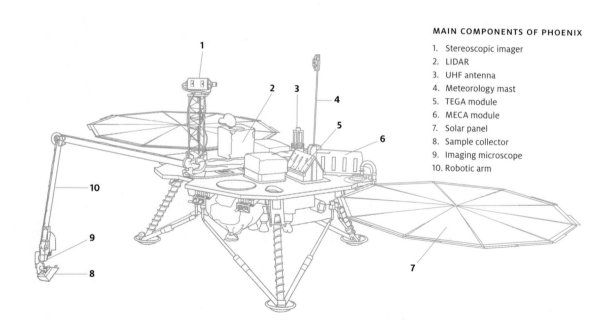

MAIN COMPONENTS OF PHOENIX

1. Stereoscopic imager
2. LIDAR
3. UHF antenna
4. Meteorology mast
5. TEGA module
6. MECA module
7. Solar panel
8. Sample collector
9. Imaging microscope
10. Robotic arm

LEFT
Installation of Phoenix into its protective nose cone atop a Delta II rocket, in preparation for launch.

RIGHT
Launch of Phoenix from Cape Canaveral in Florida on August 4, 2007.

BELOW
Inspection of Phoenix' circular solar panels at the Lockheed Martin Space Systems plant in Denver in September 2006.

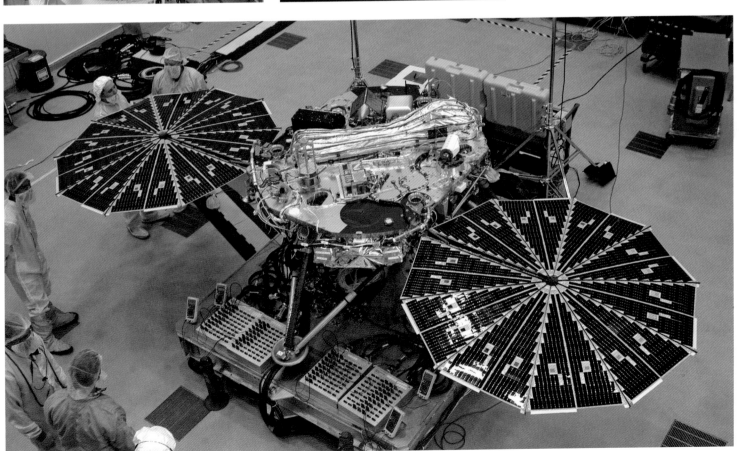

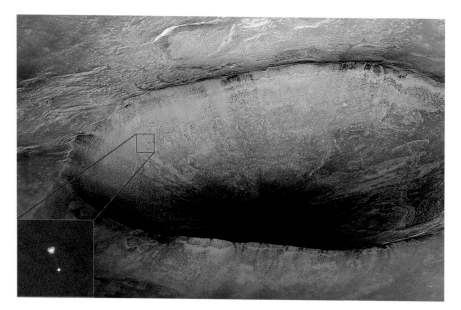

LEFT

An incredible technological first: thanks to its HiRISE camera, the Mars Reconnaissance Orbiter was able to photograph the Phoenix lander during its parachute descent on Mars.

BELOW

Geographic location of the Phoenix lander on an arctic plain, at 68.2° N latitude and 125.7° W longitude.

OPPOSITE PAGE, TOP

Artist's impression of the lander, with its retrorockets firing a fraction of a second before touchdown.

OPPOSITE PAGE, BOTTOM

Artist's impression of the Phoenix lander getting ready to dig a trench and analyze the arctic soil on Mars.

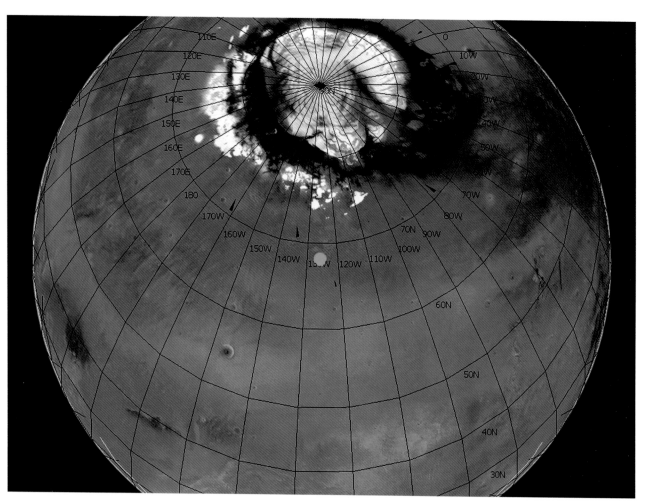

The first image of Martian polar soil sent by Phoenix.

A soil sample is filtered for entry into the microscope window of the MECA analytical lab on Phoenix. The MEGA and TEGA results indicated the unexpected presence of perchlorates in the Martian soil.

various international partners, foremost among them the Canadian Space Agency. The mission had two objectives: to study the past history of water on Mars and to look for evidence that underground ice might constitute an environment for microbial life. To these ends, the Phoenix lander was equipped with a battery of instruments, including a robotic arm capable of scraping the soil to reach permafrost ice and collect samples for physical and chemical analysis on board. For the first time in history, Martian water was actually "touched" directly. The TEGA (Thermal and Evolved Gas Analyzer) consisted of eight furnaces for the purpose of volatilizing soil components by gradually heating them to 1,000°C (1,832°F). The ovens are linked to mass spectrometers to identify the molecules released. No one had tried to detect biological molecules since the efforts of the Viking landers, 30 years before. Since any contamination from Earth could affect the results, prior Mars landers had been subjected to extensive sterilization procedures. With Phoenix, NASA dictated that additional precautions be taken with the 2.5-meter robotic arm that would contact with the soil: sterilization had to be assured to no more than one bacterial spore per square meter. Soil samples were also analyzed in the MECA module (Microscopy, Electrochemistry and Conductivity Analyzer). This module was designed for examination of soil particles by "classical" light microscopy and an atomic force microscope (the first on a space probe) to provide very high magnification — 20 times that of the optical microscope. MECA also served as a wet lab to sample liquids for biologically important properties like pH, salinity and potential for oxidation. Naturally the lander was also equipped with a stereoscopic camera to examine both the near and far surroundings, and a sophisticated weather station was provided by the Canadian Space Agency, with a LIDAR analyzer to study clouds and dust suspended in the Martian atmosphere. Once on Mars, Phoenix depended on the existing orbiters (Mars Odyssey, Mars Reconnaissance Orbiter and Mars Express) to relay UHF communications with Earth. Compatibility of communications signals was assured through use of the international protocol, Proximity-1, which had recently been established.

The turbulence caused by the Phoenix retrorockets dislodged a chunk of ice that was buried just beneath the surface, nicknamed the "Snow Queen."

LEFT

A small DVD from the Planetary Society, visible at left on the Phoenix platform, contains greetings for future Martian explorers, as well as several works of science fiction and the names of more than 250,000 earthlings.

The scoop at the tip of the robotic arm made it possible to dig beneath the surface. The soil was more compact and cohesive at this location than anticipated, making it very difficult to transfer samples into the analysis chambers.

RIGHT
This self-portrait of Phoenix, taken by its stereoscopic camera from overhead, was compiled from more than 500 images taken between June 5 and July 12, 2008. North is at the top in this picture.

OPPOSITE PAGE
This picture of the Phoenix landing site shows a polygonal pattern caused by alternate freezing and thawing, cryo-patterns similar to those found in the Canadian Arctic.

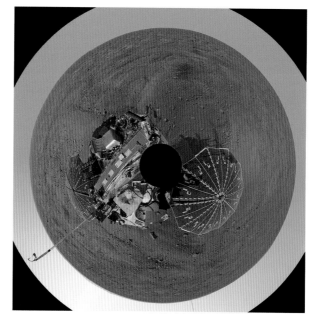

Phoenix was launched on August 4, 2007 by a reliable Delta 2925 rocket, formerly known as Delta II 7925, which had previously launched several interplanetary probes. After an extended 679-million kilometer trip, which took it around the Sun and lasted 10 months, Phoenix entered the Martian atmosphere at the high velocity of 5.7 kilometers per second. It landed directly on the plain Vastitas Borealis, close to the north polar cap, on May 26, 2008. Unlike Pathfinder and the two rovers, the Phoenix landing was achieved only with a parachute and retrorockets. To save weight, no inflatable balloons were used to soften the shock at the last moment. The last Martian landings to use this audacious method were the Viking missions, back in 1976!

Originally a different landing site was selected for Phoenix, based on images provided from orbit by Mars Global Surveyor and Mars Odyssey. The low inclination of the site and relatively few rocks seemed well suited for a landing, but more recent, very high-resolution images from Mars Reconnaissance Orbiter revealed numerous smaller rocks that could have destabilized the Phoenix lander. Taking this new information into account, the flat and monotonous Vastitas Borealis site was chosen instead. Data from Mars Odyssey had shown large quantities of shallow ice deposits (5 to 10 centimeters deep) located there. Phoenix detected water ice and discovered the presence of perchlorates in the soil, but did not report any trace of organic compounds. The last communication between Phoenix and its relay, Mars Odyssey, took place on November 2, 2008. After trying to reestablish communication on November 11, mission control declared it over on November 29. Phoenix had succumbed to the Martian winter, which progressively enveloped the probe and its solar panels in a thick layer of frozen carbon dioxide. ♂

The Giant Planets

F OUR GIANT PLANETS reside beyond Mars and the asteroid belt: Jupiter, Saturn, Uranus and Neptune. Located in a vast area 5 to 30 times the distance between the Earth and Sun, they dominate the solar system both by their mass and their gravitational force. In comparison, the small rocky planets near the Sun, including Earth, appear like insignificant vestiges of the original protoplanetary nebula.

EARTH

	JUPITER	SATURN	URANUS	NEPTUNE
Equatorial diameter (km)	142,984	120,536	51,118	49,922
Mass (Earth = 1)	318	95	14,5	17
Volume (Earth = 1)	1321	764	63	58
Average distance from the sun (UA)	5.2	9.6	19	30
Rotation period	9 h 55 min	10 h 45 min	17 h 14 min	16 h 05 min
Orbital period (years)	11.9	29.66	84.3	164.3
Moons	≥ 63	≥ 60	≥ 27	≥ 13

A new red spot on Jupiter, nicknamed Junior, was detected by the Hubble Space Telescope in April 2006.

Jupiter ♃

Jupiter is the fifth and the most massive planet in the solar system. Its mass equals more than twice that of all other planets together. Because it is the third brightest object in the night sky after the Moon and Venus, it has been known to humans from ancient times. The Romans named it in honor of their mighty god of all other gods. It forms part of the four gas giants of the solar system known as the Jovian planets, which are all composed of a small rocky core surrounded by a thick atmosphere composed mainly of hydrogen and helium. Jupiter would have had to have been even more massive to initiate nuclear fusion and thus become a star, but nevertheless it emits more energy than it receives from the Sun. This is caused by an internal compression mechanism (the Kelvin-Helmholtz mechanism), which forces Jupiter to contract and generate heat.

In 1610, Galileo (1564–1642) observed changing color patterns on Jupiter and its four largest moons: Io, Europa, Ganymede and Callisto (known as the Galilean moons). All of Jupiter's 16 main satellites were named after the mythological conquests of Zeus, the equivalent of the god Jupiter in ancient Greece. The Great Red Spot, a gigantic Jovian storm that has been observed for many centuries, was discovered by Cassini in 1665.

The discovery of moons orbiting Jupiter cast further doubt upon the geocentric theory of the Universe and supported the alternate heliocentric theory proposed by Nicolas Copernicus (1473–1543), and which caused Galileo so much difficulty during the Inquisition (see page 237).

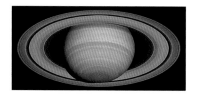

Saturn showing its South Pole and fully inclined rings — a delightful sight for astronomers.

Saturn ♄

Saturn is the sixth planet in the solar system and the second most massive. Its characteristic rings, readily observed with a small telescope, make this planet one of the most spectacular celestial objects. Galileo was the first to observe the rings in 1610, but due to their variable appearance and the poor quality of his telescope, he did not discern their proper structure, thinking they represented several orbiting moons. Saturn has several small moons less than 50 kilometers in diameter, but also has a most imposing moon, Titan, appropriately named since it is larger than either Mercury or Pluto. Titan has the unusual distinction of being the only moon in the solar system to have a dense atmosphere.

Uranus ⛢

Unknown to early astronomers, the seventh planet, Uranus, was accidentally discovered in 1781 by William Herschel, an English astronomer of German origin. Viewed through a telescope, Uranus appears as a light blue-green disc without distinguishing features; the color is due to the presence of methane in the atmosphere. Inclined at 98 degrees to the plane of its orbit, the planet gives the impression of rolling on its axis. The rings of Uranus were discovered in 1977 while it occulted a star. Unlike other giant planets, Uranus does not have a source of internal heat; it can only capture thermal energy from the Sun. At 19 times farther away from the Sun than Earth, it receives nearly 400 times less sunlight. It has only been visited by one space probe, Voyager 2, which flew by in 1986 and obtained the first detailed images of its five major moons: Titania, Oberon, Umbriel, Miranda and Ariel.

William Herschel (1738–1822)

Neptune ♆

During the half-century that followed the discovery of Uranus, astronomers tried in vain to accurately predict its orbit. This gave rise to the idea that a massive planet farther away from the Sun was affecting its trajectory, and in light of Newton's laws of gravity, the challenge of the search became a topic of interest. In 1845–46, the young English mathematician John Couch Adams (1819–1892) and the French astronomer Urbain Le Verrier (1811–1877) independently calculated the position of this putative eighth planet. Shortly after, on September 25, 1846, its existence was confirmed by the Berlin Observatory. The blue-green color of the planet inspired the name of Neptune, the Roman god of the seas. Similar in shape and composition to its calmer brother, Uranus, Neptune displays frequent changes due to violent winds that displace clouds of methane crystals at speeds of more than 1,000 kilometers per hour. Unlike Jupiter's Great Red Spot, which had been observed for many centuries, a similar phenomenon on Neptune, the Great Dark Spot, disappeared a few years after the flyby of Voyager 2 in August 1989 (the planet's only visit by a space probe). Neptune's largest moon, Triton, was discovered in 1846 by the English astronomer William Lassell (1799–1880), less than three weeks after Neptune was confirmed. According to the data obtained from Voyager 2, Triton has an extremely cold surface (–239°C), but nevertheless is geologically active, with cryo-volcanoes (ice volcanoes) that eject nitrogen or liquid-methane jets several kilometers in height.

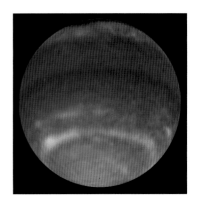

Although it is the coldest planet in the solar system, color variations in its cloud bands indicate that Neptune undergoes seasonal changes.

Pioneer 10 and 11

BELOW, LEFT

Pioneer 10 is shown secured to its launch adapter, before encapsulation into the rocket's nose cone at the Kennedy Space Center in Florida.

BELOW, RIGHT

Launched by an Atlas-Centaur rocket on March 2, 1972, Pioneer 10 became the first space exploration mission to the asteroid belt and Jupiter.

OPPOSITE PAGE

A replica of Pioneer 10 on display at the National Air and Space Museum, Washington, D.C.

A<small>T THE END OF THE</small> 1960<small>S</small>, the greatest challenge for solar-system exploration was a visit to Jupiter, Uranus and Neptune. Astronomers calculated that a "Grand Tour" was possible, since the four giant planets would be aligned on the same side of the Sun some time in 1977–78. Since astronomical distances in the order of billions of kilometers had to be traveled, engineers were tempted to come up with new propulsion modes to reach greater speeds. Since the innovative ion-drive and nuclear propulsion modes were not quite up to the task, they were still dependent on chemical propulsion systems. An economical solution involved adapting existing spacecraft for much longer trips to the giant planets. The Pioneer series of probes met that need, since Pioneer 6 to 9 had successfully navigated heliocentric orbits.

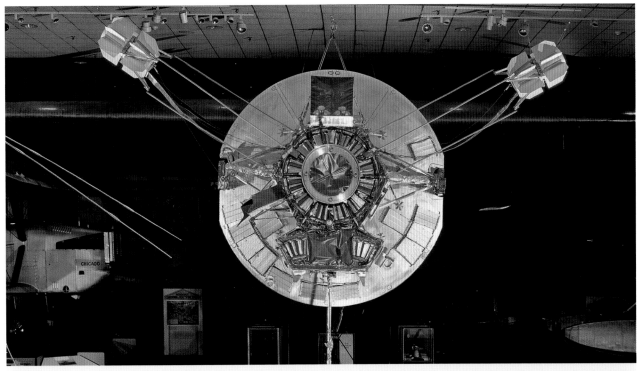

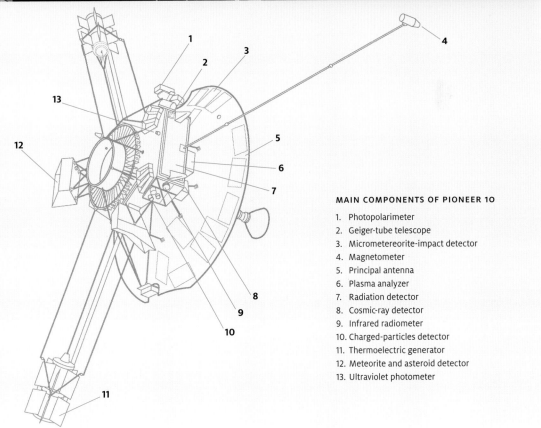

MAIN COMPONENTS OF PIONEER 10

1. Photopolarimeter
2. Geiger-tube telescope
3. Micrometereorite-impact detector
4. Magnetometer
5. Principal antenna
6. Plasma analyzer
7. Radiation detector
8. Cosmic-ray detector
9. Infrared radiometer
10. Charged-particles detector
11. Thermoelectric generator
12. Meteorite and asteroid detector
13. Ultraviolet photometer

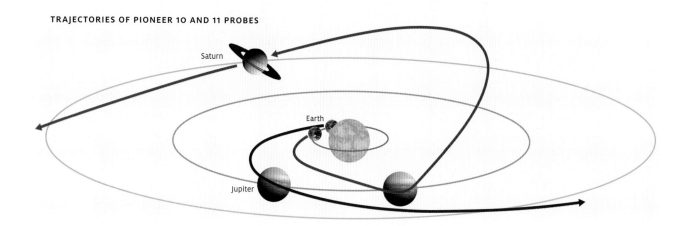

Saturn

Earth

Jupiter

In 1969, a company in California called TRW (now Northrop Grumman) was awarded the construction contract for two Pioneer probes of the "Jupiter" category, as well as another probe in reserve. Coordination of the program was under the direction of the NASA's Ames Research Laboratory, near San Francisco. The Atlas-Centaur rocket, the most powerful after the Saturn 1, was chosen to send the probe (285 kilograms and 2.9 meters tall) on a direct trajectory toward Jupiter. Since Jupiter is too far from the Sun for regular size solar panels to be effective, electric power for Pioneer 10 and 11 was generated by thermoelectric generators containing Plutonium-238 pellets. The 150-watt generators were mounted on two girders, separated by a distance equal to the length of the probe itself, so that the radiation did not interfere with the instrumentation. The 25-kilogram payload of scientific equipment aboard Pioneer 10 (11 instruments in total) consisted of an imaging photopolarimeter and several sensors to analyze thermal and ultraviolet radiation, magnetic fields, solar wind, cosmic rays, and micrometeorite impacts. Image capture was based on an original sweeping technique. While the probe was stabilized by constantly spinning about its axis, the photopolarimeter camera line-scanned the target with each rotation (the spin-scan method). This way hundreds of scans were accumulated and then the computer assembled them into a complete image. Pioneer 10 lifted off from Cape Canaveral on March 3, 1972. A huge unknown during this journey was the potential danger of collision when first crossing the asteroid belt, between the orbits of Mars and Jupiter.

On July 15 1972, Pioneer 10 became the first space probe to enter the asteroid belt. Although it detected a higher density of dust in this area, it passed through without trouble and without brushing against any asteroids. When Pioneer 10 was 25 million km from Jupiter on November 6, it obtained the very first picture of the giant planet taken from space, and on December 3, 1973, it transmitted the first close-up images of Jupiter, while flying by the planet at a distance of 130,000 kilometers. The clarity of these images greatly surpassed those obtained with the Earth-based telescopes of the time, and revealed strong atmospheric turbulences caused by thermal energy from Jupiter's core. The detection of a strong magnetic field around Jupiter also provided much information about the planet's core.

By crossing the orbit of Neptune on June 13, 1983, this American spacecraft became the first human-made object to venture beyond the planetary zone of the solar system. After the probe's mission ended on March 31, 1997, scientists in charge of Pioneer 10 decided to continue tracing its weak signal to help train operators of the Deep Space Network radio telescope array. A final clear telemetry session lasting 33 minutes was received on April 27, 2002. This remarkable spacecraft transmitted its

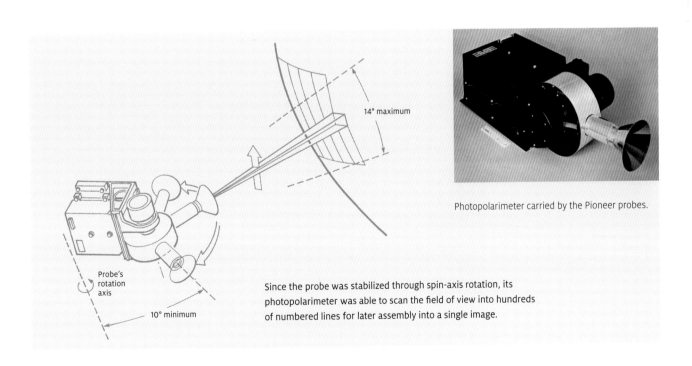

14° maximum

Probe's rotation axis

10° minimum

Photopolarimeter carried by the Pioneer probes.

Since the probe was stabilized through spin-axis rotation, its photopolarimeter was able to scan the field of view into hundreds of numbered lines for later assembly into a single image.

First close-up view of the Great Red Spot, obtained on December 4, 1974, by Pioneer 11 during its flyby, 540,000 kilometers above the cloudy surface of Jupiter. The smallest visible object measures approximately 250 kilometers.

last signal on June 22, 2003, at a distance of 12.5 billion kilometers from Earth (or 82 times the Earth–Sun distance). Its radio signal, operating at a mere 7.5 watts or the equivalent of a Christmas tree bulb, was only a millionth of a billionth of a watt by the time it reached one of the three 64-meter antennas of the Deep Space Network. Even traveling at the speed of light, it took the signal 11 hours and 20 minutes to reach Earth. In 2011, Pioneer 10 is heading toward the star Aldebaran in the constellation Taurus. At its current speed of 2.6 AU annually (12.5 kilometers per second), it will need 2 million years before reaching it ...

It was not possible for Pioneer 10 to approach Saturn after its meeting with Jupiter. Pioneer 11 became the first mission in history to attempt and succeed at a flyby of Saturn, before also leaving the solar system. Pioneer 11 left Earth on April 6, 1973, on an Atlas-Centaur rocket and reached Jupiter after a 20-month voyage. On December 4, 1974, it flew by the planet only 34,000 kilometers above the upper cloud layer and transmitted spectacular images of both the Great Red Spot and the planet's polar regions (which had never been seen from Earth). Pioneer 11 flew by Saturn on September 1, 1979, at a distance of 21,000 kilometers above its atmosphere. The probe nearly collided with a heretofore unknown moon, Epimetheus, missing it by only a few thousand kilometers. Pioneer 11 imaged the moons Janus and Mimas, discovered additional rings around Saturn, and reported that the mysterious Titan was too cold to sustain life as we know it on Earth. As Pioneer 11 was flying over Saturn, the two Voyager probes had already passed Jupiter. In order to live up to its pioneering name, it was decided to make Pioneer 11 pass through Saturn's rings and follow the trajectory planned for the Voyager probes. In case of an accident, Voyager missions and trajectories would be modified accordingly. However, this might mean that it would not be possible for Voyager to approach Saturn close enough to benefit from the required gravitational boost to reach Uranus and Neptune. On September 1, 1979, Pioneer 11 survived its passage through the rings without incident, in spite of Saturn's strong magnetic field. During this passage, Pioneer 11 transmitted uniquely detailed images of Saturn's atmosphere and its rings before moving away for good

ABOVE

On September 1, 1979, Pioneer 11 approached the giant planet to 22,000 kilometers, for the first historic flyby.

LEFT

Saturn, accompanied by Titan, as imaged by Pioneer 11.

BELOW

Artistic interpretation of Pioneer 10 leaving the solar system.

toward the edge of the solar system. In February 1985, the probe's nuclear-powered generators began to show serious signs of decay, and on September 30, 1995, at 6.5 billion kilometers from Earth, its daily activities ceased. The last radio signals from Pioneer 11, at the very limit of detectability, were received in November 1995. At the time of this writing, the space-craft is headed in the direction of the constellation Aquila, northwest of the constellation Sagittarius, at a speed of 2.4 AU per year (11.6 kilometers per second). If all goes well, it should approach a star in about four million years. Both Pioneer 10 and 11 carry with them a message for any extraterrestrial intelligence that might have the technology to capture and intercept an inert artificial object cruising interstellar space. The content of the message, engraved on a plaque, was the subject of controversy and varied interpretations (see following page).

A message for extraterrestrials

In a historic first, the Pioneer 10 and 11 space probes carried a message destined for an extraterrestrial civilization. The idea originated with science journalist Eric Burgess, during a visit to TRW Systems as Pioneer was undergoing testing. With NASA's approval, the message was conceived in three weeks by astronomer Carl Sagan (see page 155), in collaboration with astrophysicist Frank Drake and journalist Richard C. Hoagland, all of whom had pondered questions regarding potential communication between humans and other intelligent species in the Universe.

The message was engraved on a gold aluminum plaque (15 x 23 centimeters), which was attached to the antenna's support to protect it against interstellar dust. The message is meant to

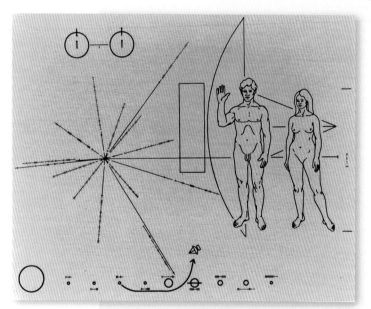

indicate where our species lives in the galaxy and at which epoch the space probes were sent. It shows a man and a woman in front of the probe's silhouette to give an idea of the scale of their dimensions. The right hand of the man is raised in a sign of a peaceful salute. The drawings are the work of Linda Salzman Sagan, who at the time was Carl Sagan's wife. The plaque also shows a representation of the hyperfine transition of hydrogen, the most abundant element in the Universe. The wavelength (21 centimeters) and the frequency (1,420 MHz) of the hyperfine transition provide the scales of length and time in the other diagrams. The values are indicated in binary notation. One diagram shows that the spacecraft comes from the third planet of a star whose position is indicated as triangulation functions of the periodicity of 14 pulsars and the distance from the Sun in the center of the galaxy (the 15 radiating dotted lines). Since pulsars change their periodicity within a known time frame, it's possible to calculate the epoch at which the probes were sent. Some critics deemed it be either in poor taste or dangerous. How would you have interpreted this universal message?

Voyager 1 and 2

THE SUCCESS OF Pioneer 10 and 11 augured well for more sophisticated future missions to the giant planets. The task fell to NASA's Jet Propulsion Laboratory Mariner program, with its Mariner 11 and 12 planetary probes. These were first renamed Mariner Jupiter-Saturn, and then Voyager, a name deemed more accessible to the media. Initially, the Voyager missions were planned as simple flyby missions to Jupiter and Saturn. Gradually the possibility of a "grand tour" was conceived that would take advantage of an exceptional alignment of the giant planets at the end of the 1970s, an occurrence that occurs only every 176 years. Using gravitational boosts, it was calculated that a Voyager probe could "ricochet" from one planet to another in order to include a close-up trip to Uranus and Neptune for the very first time. With gravitational assistance, the Earth–Neptune voyage would only take 12 years instead of the 30 years required with a conventional trajectory. That also meant less fuel and lower weight at launch time.

Voyager 1 undergoing testing in the Space Simulator at the Jet Propulsion Laboratory in Pasadena, California.

FAR LEFT
Encapsulation procedure of Voyager 2, in preparation for the launch.

LEFT
Liftoff of the interplanetary probe Voyager 1, on September 5, 1977, setting out on a very long, discovery-filled voyage.

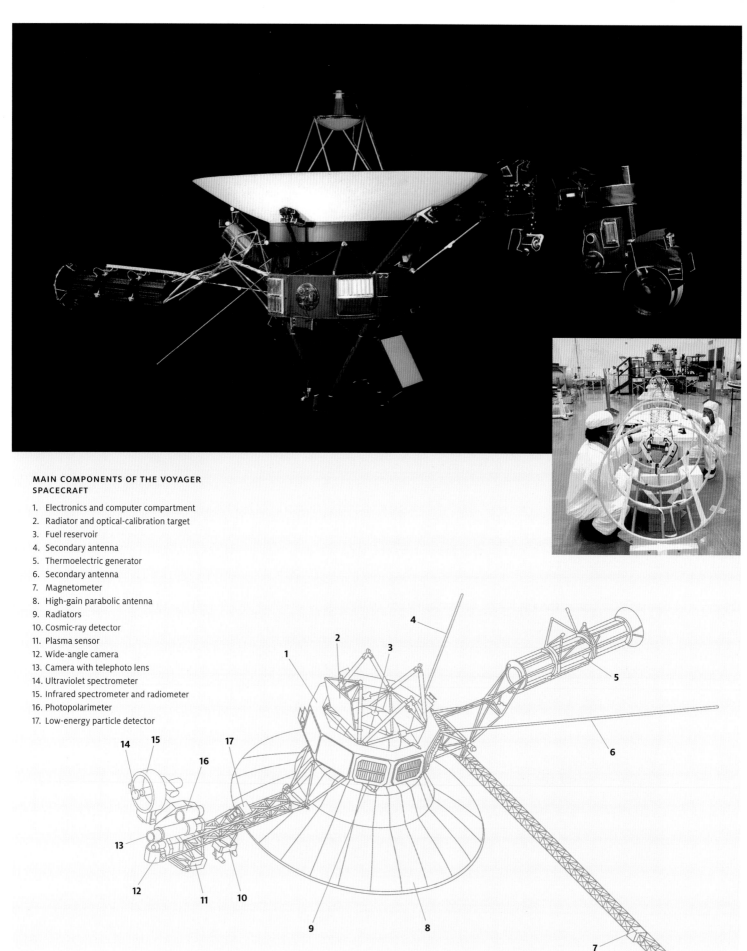

MAIN COMPONENTS OF THE VOYAGER SPACECRAFT

1. Electronics and computer compartment
2. Radiator and optical-calibration target
3. Fuel reservoir
4. Secondary antenna
5. Thermoelectric generator
6. Secondary antenna
7. Magnetometer
8. High-gain parabolic antenna
9. Radiators
10. Cosmic-ray detector
11. Plasma sensor
12. Wide-angle camera
13. Camera with telephoto lens
14. Ultraviolet spectrometer
15. Infrared spectrometer and radiometer
16. Photopolarimeter
17. Low-energy particle detector

Unlike Pioneer 10 and 11, the 815-kilogoram Voyager spacecraft was stabilized along three axes, thanks to gyroscopes and an optical navigation system, thereby pointing its high-gain antenna toward Earth. In addition to experiments with radio communications (radio science), Voyager's scientific exploratory mission relied on a panoply of instruments that included an infrared spectrometer and radiometer, an ultraviolet spectrometer, a photopolarimeter, two magnetometers, a cosmic-ray sensor, a plasma sensor and a charged-particle sensor. The imaging system was placed on a movable platform and equipped with a 200 mm f/3 wide-angle camera and a camera with a 1,500 mm f/8.5 telephoto lens for close-up imaging. The cathode-ray, vidicon-tube cameras were improved versions of the ones used on previous Mariner missions. An 8-filter wheel, mounted in front of each camera, allowed it to capture images of different wavelengths. The on-board computers were equipped with three microprocessors: military version 8-bit CMOS 1802 from RCA, rated at 6.4 MHz and hardened to resist radiation and electrostatic discharges and support wide temperature variations (−55°C to 125°C). The Voyager probes had three nuclear (plutonium-238) thermoelectric generators, capable of delivering 470 watts of electricity at launch time. Since plutonium-238 has a half-life of 87.74 years, the generator will only lose 0.78 percent of its power every year.

Voyager 2 was the first to be launched, on August 20, 1977 from Cape Canaveral by a Titan IIIE-Centaur rocket, followed by Voyager 1 on September 5. Despite this slight time delay, the latter was launched on a more direct and faster trajectory, thereby allowing it to reach Jupiter and Saturn before its twin. Voyager 1 flew by Jupiter on March 5, 1979, at a distance of 278,000 kilometers. Most of the close-up images of Jupiter, including the discovery of its thin ring system, its moons, and measurements of

OPPOSITE PAGE, TOP

An overview of a Voyager probe. The main antenna has a diameter of 3.7 meters. Deployment testing of the 13-meter fiberglass beam that supports the magnetometers.

TRAJECTORIES OF VOYAGER 1 AND 2 PROBES

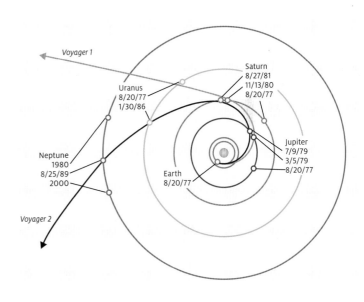

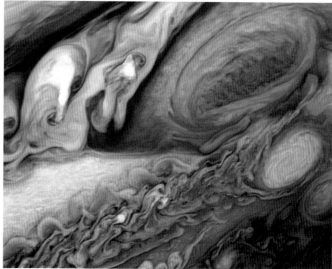

The Great Red Spot region, as imaged by Voyager 1 when it was 9.2 million kilometers from Jupiter. Detail is visible to 160 kilometers. To give a sense of scale, it should be noted that the white oval below the Great Red Spot is the diameter of Earth.

Details of the strong atmospheric turbulence in regions near the Great Red Spot. The colors shown are artificial to enhance contrast. Note the differences with the preceding image.

the planet's magnetic field and radiation belt, were obtained during the 48 hours of the flyby. Voyager 1 discovered Jupiter's moons Metis and Thebe, as well as the Saturn satellites Prometheus and Pandora, which shepherd Saturn's F ring. During a close approach to Jupiter's moon Io, the spacecraft photographed erupting sulfur volcanoes, marking the first time that volcanic activity was documented on another celestial object.

As planned, Voyager's trajectory was directed toward the planet Saturn, with the help of a gravitational boost from Jupiter. The Saturn flyby took place on November 2, 1980, at an altitude of 124,000 kilometers above the planet's clouds. Mission control had given highest priority to a close-up view of the moon Titan and its atmosphere, but such an approach deflected the probe's trajectory away from the ecliptic, so it precluded any possibility of a "Grand Tour" toward Pluto, which was considered a lesser priority.

A close-up view of Europa taken by Voyager 2, showing the characteristic smooth surface of this moon. The absence of impact craters suggest a fairly recent resurfacing.

For its part, Voyager 2 flew over Jupiter on July 9, 1979, at a distance of 570,000 kilometers. It confirmed the volcanic activity on Io and provided detailed views of the surface of Europa, showing a surprising pattern of crossed linear features, which were interpreted as a coat of ice. The spacecraft got closest to Saturn on August 26, 1981, and explored the planet's ring structure in detail, as well as the surface of several moons, including Enceladus and Iapetus. Following another gravitational boost, this time by Saturn, Voyager 2 became the first spacecraft to reach the giant planet Uranus, approaching it to within 81,500 kilometers on January 24, 1986. The probe examined the planet's rings and transmitted images of Miranda and 10 other heretofore unknown moons: Cordelia, Ophelia, Bianca, Cressida, Desdemona, Juliet, Portia, Rosalind, Belinda and Puck. Examination of Miranda's surface details showed it to be one of the oddest objects in the solar system. Voyager 2's next destination, the planet Neptune, was reached on August 25, 1989, another historic first. A scientific surprise was in store when it flew by the planet's giant moon Triton. Several geysers were discovered, which spewed nitrogen gas and formed a very tenuous atmosphere around Triton. The pressure is only 1/70,000th that of Earth's atmosphere. The information obtained about Triton suggests that this moon has much in common with Pluto and, like the latter, was probably a Kuiper Belt object captured by Neptune's gravitational field. Voyager 2 discovered and additional six additional moons around Neptune: Naiad, Thalassa, Despina, Galatea, Larissa and Proteus.

BELOW, LEFT
One of the great achievements of the Voyager missions was the discovery of active volcanoes on Jupiter's moon, Io. The gas plumes can rise up to 300 kilometers in height.

BELOW, RIGHT
A scaled photomontage of the four Galilean moons and the small moon, Amalthea, all of which were imaged by the Voyager probes during their flyby of Jupiter. Shown in their traditional order are Io, Europa, Ganymede and Callisto.

Saturn accompanied by the two moons, Tethys and Dione, as seen by Voyager 1 on November 3, 1980, from a distance of 13 million kilometers. The shadow of the moon Tethys is visible under the A and B rings, which are separated by Cassini's Division.

In this digitally enhanced Voyager 2 image, Saturn's rings can be differentiated by the size of their particulates and their chemical composition.

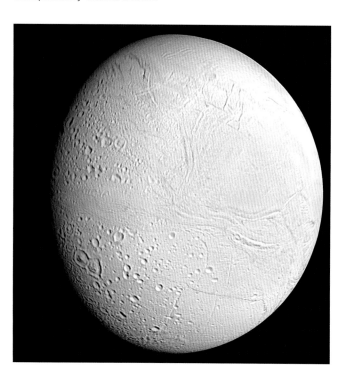

The small moon Mimas was struck by a massive object that left a disproportionably large impact crater, named Herschel.

Voyager 1 was the first to measure the thickness of Titan's atmosphere, which is much denser than that of Mars. Its bluish color is due to the presence of methane gas.

Image of the moon Enceladus, as seen by Voyager 2. This 500-kilometer diameter moon, covered in ice, is one of the most volcanically active of the solar system. The smallest visible feature on this image measures 1 kilometer.

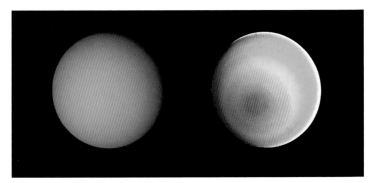

The first encounter with Uranus, on January 10, 1986, when Voyager 2 was about 18 million kilometers from the planet. The bluish tint of the atmosphere is due to the presence of methane. In the image on the right, colors have been artificially enhanced to reveal the planet's rotational axis (orange area), which is inclined to the plane of the ecliptic.

The nine rings of Uranus, captured by Voyager 2's camera: Epsilon, Delta, Gamma, Eta, Beta, Alpha, 4, 5 and 6.

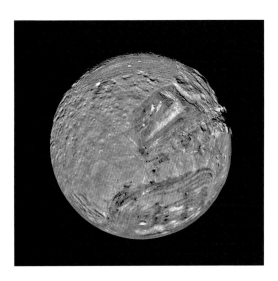

The moon Miranda, with its mosaic surface features that speak to its violent past. Some of its canyons reach a depth of 20 kilometers.

The Voyager probes are now in their interstellar phase. Voyager 1 is presently the furthest human-made object from the Sun and has already passed the astonishing distance of 100 AU, or 15 billion kilometers. In less than one day, on December 16, 2004, Voyager 1 achieved another milestone by crossing the terminal shock, the turbulent outer shell of the sun's sphere of influence, called the heliosheath. That is a transitional region, where the strength of the solar wind diminishes suddenly, going from 700 kilometers per second to subsonic speeds. On December 13, 2010, NASA announced that Voyager 1 had reached the confines of our solar system and was now in a region where there is no outward motion of solar particles. In 2015, it will enter the heliopause, where solar particles give way to the interstellar medium. The space probe is leaving us at a speed of 17 kilometers per second. In 2011, 34 years after it was launched, the spacecraft's electric generating capacity is theoretically down to 75 percent, because of radioactive decay of the plutonium-238 power source. In practice, however, this only provides 60 percent of the probe's maximum power, since the thermoelectric couplers, which convert heat into electricity, have also degraded. Although the probe itself remains operational, some system redundancies have been eliminated to conserve energy. When the spacecraft leaves the heliosphere beyond heliopause and moves into interstellar space, it will enter a region beyond the Sun's influence and travel into the realm of our galaxy. For its part, Voyager 2 will cross the 100 AU distance threshold in 2012. The nuclear generators of both probes will provide enough energy to fuel them until

The first close-up image of Neptune taken on August 20, 1989, by Voyager 2. The discovery of Neptune's turbulent atmosphere indicates that the planet has an internal energy source.

at least the year 2025. After this, they will quietly continue along their respective trajectories. Voyager 1 is going in the direction of the star AC+793888, in the constellation of Cameleopardalis (the Giraffe), and will reach it in about 40,000 years, while Voyager 2 is heading toward the brightest star in our sky, Sirius, in the constellation Canis Major, reaching it in 296,000 years. As with Pioneer 10 and 11, the two Voyagers carry a message, this time in the form of a digital record. This is intended for any extraterrestrial civilization that is advanced enough to intercept the probes (see page 228).

The Voyager program can be considered a remarkable scientific and technical success with a unique place in NASA's history. It will remain so for quite some time, due to its record exploration of 48 moons and four planets, including two that had never been visited before — Uranus and Neptune. As if this abundance of data was not enough, the Voyager probes are still sending information more than 30 years later, on the frontiers of the solar system!

Triton, Neptune's largest moon, is one of three known worlds, along with Earth and Titan, that has nitrogen in its atmosphere. Voyager 2 identified cryo-volcanoes on Triton that spew plumes of liquid nitrogen, methane and dust. Triton's temperature is so low (–238°C) that its surface is covered with solid nitrogen and methane ice.

Leaving the solar system, Voyager 1 entered a new frontier in December 2004, by crossing the terminal shock of the heliosphere, located at 94 AU from the Sun. Voyager 1 is now in the interstellar phase of its mission. Voyager 2 crossed the terminal shock in August 2007 at 84 AU, showing that the heliosphere is an irregular structure, due to the influence of galactic magnetic fields.

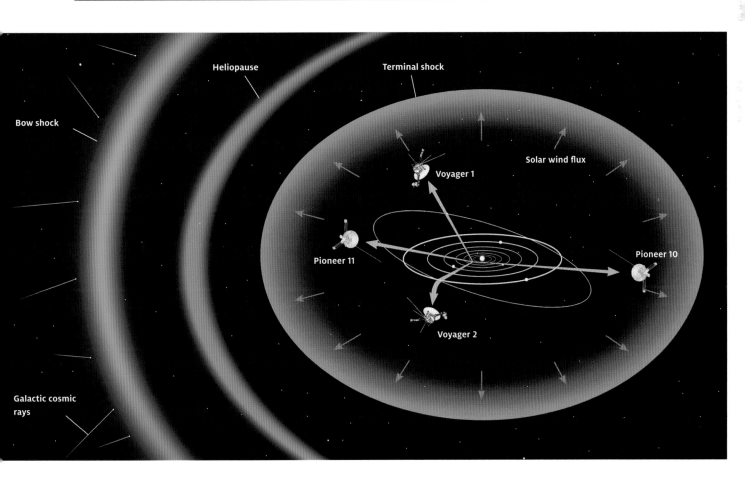

Heliopause

Terminal shock

Bow shock

Voyager 1

Solar wind flux

Pioneer 11

Pioneer 10

Voyager 2

Galactic cosmic rays

Interstellar time capsule

There is a chance, however extremely remote, that one of the Voyager probes now leaving the solar system could one day be intercepted by an extraterrestrial civilization (or by humans in the distant future). They will discover that the probe carries a message, more sophisticated than the simple plaque of the Pioneer probes (see page 218). The message is recorded on a gold-coated copper phonograph disk and contains sounds and pictures representing the diversity of life and humanity on Earth. The cover provides diagrammatic instructions on how to use the enclosed cartridge and needle in order to read the disk's analog audio and image files. Diagrams indicating where the spacecraft came from in the galaxy and a fragment of pure uranium-238 are also included to allow the reader to date its construction. The physicist Carl Sagan (see page 155) headed the committee charged with the selection of the sounds and images. The sound portion of the disc contains "Hello to all" in 55 languages. Following this, a section called "Earth Sounds" consists of various natural sounds and a 90-minute eclectic selection of oriental and western classical music. The 115 images consist of diagrams and photographs of our planet, our anatomy and examples of our science and technology.

"This is a present from a small, distant world, a token of our sounds, our science, our images, our music, our thoughts and our feelings. We are attempting to survive our time so we may live into yours."

—excerpt from the Voyager message from U.S. President Jimmy Carter

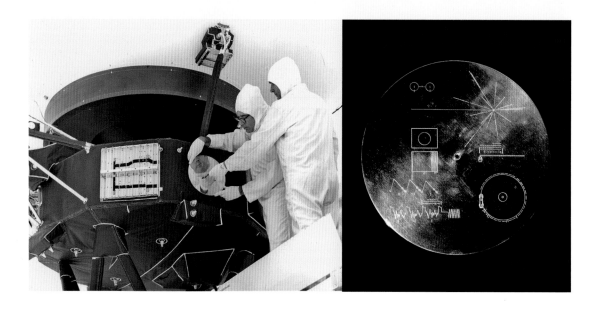

Galileo

THE LARGE QUANTITY OF scientific information obtained during brief flybys of Jupiter by Pioneer and Voyager fully justified sending an orbiting spacecraft around the giant planet for a multi-year mission. This project, christened Galileo in honor of the famous Italian astronomer and physicist (see page 237), was being developed even before the Voyager probes left to explore the outer solar system in 1977. The preparation of the Galileo mission was a long and frustrating undertaking for those involved.

BELOW, LEFT
The Galileo probe being coupled to the IUS booster rocket.

BELOW, RIGHT
Like Magellan and Ulysses, Galileo was one of the space probes lifted into low Earth orbit by the space shuttle, in this case, Atlantis, during the mission STS-34. Liftoff happened on October 18, 1989, before it was injected into a transfer orbit on its way to Jupiter.

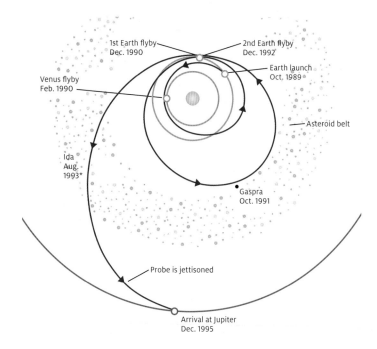

1st Earth flyby
Dec. 1990

2nd Earth flyby
Dec. 1992

Earth launch
Oct. 1989

Venus flyby
Feb. 1990

Asteroid belt

Ida
Aug.
1993*

Gaspra
Oct. 1991

Probe is jettisoned

Arrival at Jupiter
Dec. 1995

Galileo's indirect trajectory toward Jupiter required several gravitational boosts.

OPPOSITE PAGE, TOP

An artist's pre-launch rendition of Galileo orbiting the Jovian system. Unfortunately, a mechanical problem during the deployment of the main antenna forced the mission to use the low-gain secondary antenna for communicating with Earth.

The initial plan was that it would be launched by the space shuttle in 1981. However, development of the shuttle was delayed and the Galileo launch was pushed back to 1984. That date was missed because of uncertainties about the type of booster needed to propel the probe once in Earth orbit. A new date in 1986 was chosen, but the Challenger shuttle explosion that year further postponed the launch. Finally, on October 18, 1989, Galileo took off aboard the shuttle Atlantis during the STS-34 mission. The Challenger disaster forced NASA to revisit the Galileo booster power for security reasons, and the less powerful IUS (Inertial Upper Stage) booster, which was also used for the Magellan mission (see page 118), was selected instead. That meant a longer voyage, requiring several gravitational boosts before reaching Jupiter, including one from Venus (February 10, 1990) and two from Earth (December 8, 1990 and December 8, 1992), through what was called a VEEGA (Venus-Earth-Earth Gravity Assist) trajectory. Galileo weighed 1,138 kilograms, including 103 kilograms of scientific equipment, and the voyage lasted six years. The length of the trip left plenty of time for technicians to test the probe's systems. During one passage, where Galileo was 960 kilometers above Earth, it took nearly 3,000 images (mainly of Australia and Antarctica), as well as photographing the Earth rotating on its own axis for the very first time. A technical glitch occurred in 1991: the umbrella-shaped high-gain radio antenna only partially deployed, rendering it unable to transmit at a high rate (134 kilobits per second). Despite all efforts to fix this mechanical problem, mission personnel were forced to use a secondary low-gain antenna, which could only transmit at the dismal rate of 160 bits per second! This called for a complete reprogramming of the on-board computers and an adjustment of the Deep Space Network antennas. In October 1995, a few weeks before the probe was to orbit Jupiter, another serious problem arose, which monopolized the attention of technicians.

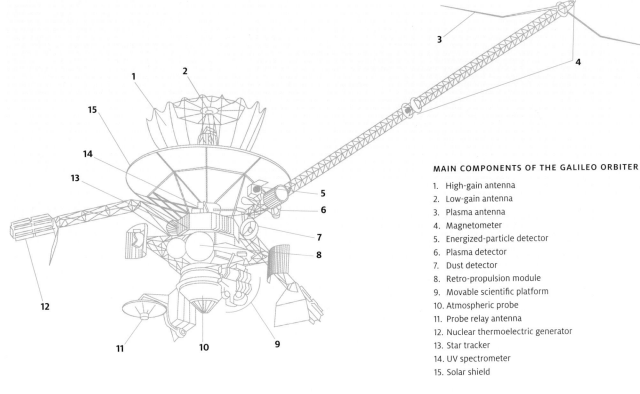

MAIN COMPONENTS OF THE GALILEO ORBITER

1. High-gain antenna
2. Low-gain antenna
3. Plasma antenna
4. Magnetometer
5. Energized-particle detector
6. Plasma detector
7. Dust detector
8. Retro-propulsion module
9. Movable scientific platform
10. Atmospheric probe
11. Probe relay antenna
12. Nuclear thermoelectric generator
13. Star tracker
14. UV spectrometer
15. Solar shield

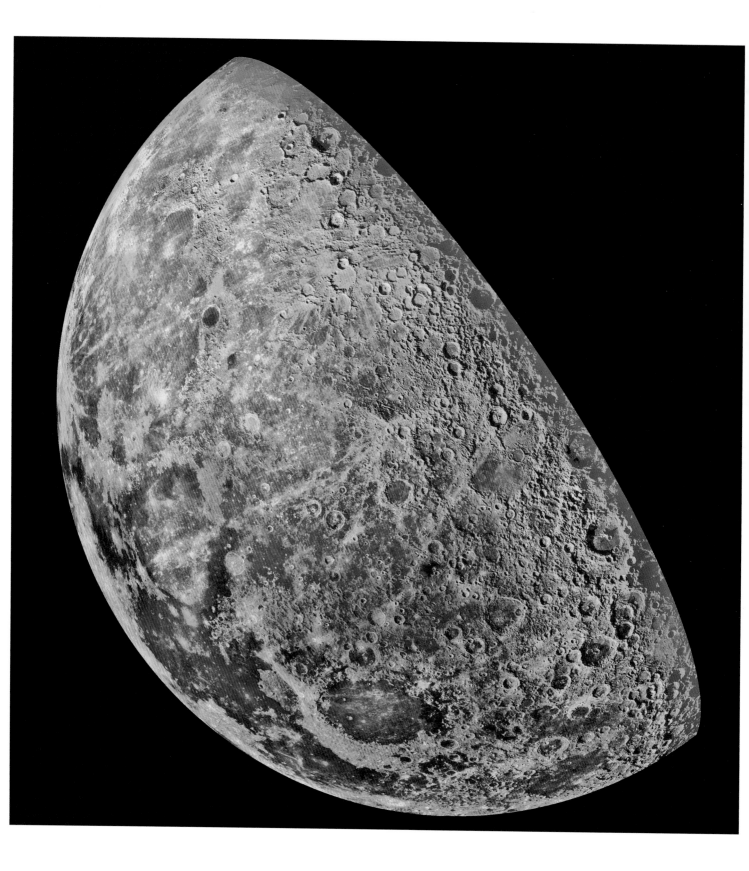

Galileo's tape recorder, with a capacity of 114 megabytes, was jammed in the rewind position for a period of 15 hours. As a result, a portion of the magnetic tape was useless, making it impossible to record data acquired on Io and Europa during the approach phase. In addition, the probe was repeatedly subjected to high doses of radiation close to Jupiter, which greatly damaged the camera and several electronic circuits.

The Galileo orbiter was structured in a dual-spin manner, with a moveable rotating component (three turns per minute) and a fixed component. The latter housed a digital imaging system, with an 800 x 800 pixel CCD camera and a 1,500 mm Cassegrain-type telescope. The image resolution was 20 to 1,000 times better than that of the Voyager missions, with better color rendition. Infrared and ultraviolet spectrometers were also located on the fixed module. All other instruments were part of the moveable module, including a magnetometer, a plasma sensor, a cosmic-dust sensor and a charged-particle sensor. A circular solar shield, located between the high-gain antenna and the probe's body, protected it from intense heat during its excursion into the inner solar system. Galileo's propulsion system, with its hydrazine engines, was the result of a collaboration between the German company DASA (Daimler Benz Aerospace AG, now integrated into the European group EADS: European Aeronautic Defense and Space Company) and NASA. Electricity was provided by two plutonium-238 thermoelectric generators, capable of generating 495 watts once the probe was orbiting Jupiter. The on-board computers used RCA's 8-bit 1802 microprocessors, as did the computers on Viking and Voyager. Galileo also had an optical star-tracking system for attitude control and pinpointing the location of the spacecraft in space.

A family portrait of Jupiter with its Galilean moons. From top to bottom: Io, Europa, Ganymede and Callisto.

OPPOSITE PAGE
Our Moon, imaged by Galileo during a gravitation-assisted maneuver. This artificially colored mosaic, constructed from 53 images taken through 3 narrow band-pass filters, shows the different rock compositions. The intense blue of Mare Tranquillitatis indicates strong concentrations of titanium.

RIGHT

This image of Io was obtained during Galileo's ninth orbit around Jupiter on July 28, 1997. The volcanic eruption photographed on the moon's left edge reached a height of 140 kilometers. Another eruption was observed near the terminator. This second volcano appears to have been active since Voyager's passage in 1979.

FAR RIGHT

Europa, seen here in false color, appears to be covered by a constantly regenerating ice mantle. The observation of changing streaks across the surface suggests that a liquid ocean may be present below the moon's icy surface.

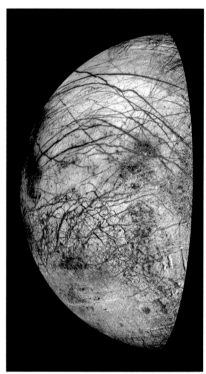

Galileo was able to approach two asteroids, Gaspra (October 29, 1991) and Ida (August 28, 1993), while crossing the asteroid belt between Mars and Jupiter (see pages 325 and 365). As luck would have it, Galileo had a ringside seat in 1994 to observe the collision of comet Shoemaker-Levy's fragments with Jupiter, while earth-based telescopes had to wait for the planet to rotate before witnessing the impacts.

Galileo carried a 339-kilogram capsule to probe the interior of the Jovian atmosphere. The probe was released toward Jupiter on July 13, 1995, and entered the atmosphere on December 7, 1995, at a speed of 48 kilometers per second. Atmospheric friction slowed it to subsonic speeds, thereby subjecting it to a gravitational force of 230 g. After the probe's 152-kilogram heat shield was vaporized through friction in a few minutes, its parachute was deployed to slow the descent, in order to measure the temperature, pressure and chemical composition of Jupiter's atmosphere. Up to 58 minutes of data were transmitted and relayed to Earth by the orbiting Galileo. Winds of up to 900 kilometers per hour were recorded and lightning was observed. The capsule sank to a depth of 150 kilometers before succumbing to extreme conditions of pressure (23 bars) and temperature (153°C). For the very first time, Jupiter's atmosphere had been studied in situ.

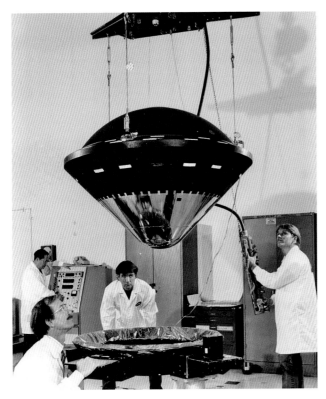

The Galileo mission's atmospheric probe faced the most difficult of all attempted entries to date. The decelerating action of Jupiter's clouds reduced the probe's speed from 48 kilometers per second to subsonic speeds in less than two minutes, thereby subjecting it to a gravitational force of 230 g!

Protected by a heat shield designed to be jettisoned, the descent module carried several instruments to measure the physical properties and chemical composition of the Jovian atmosphere at various altitudes.

MAIN COMPONENTS OF GALILEO'S ATMOSPHERIC PROBE

1. Antenna
2. Parachute compartment (empty)
3. Lightning sensor
4. Antistatic device
5. Nephelometer
6. Stabilizing fin

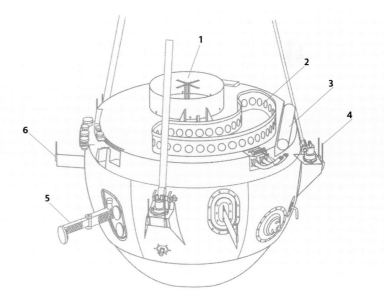

Details of the Great Red Spot with artificial coloring. This mosaic was assembled from 18 images recorded through filters in the near-infrared. The colors indicate the relative altitude of clouds (pink: more elevated, blue: less elevated).

Jupiter's main ring and its halo. The halo is formed by a cloud of small electrically-charged particles emitted by the rings and displaced by Jupiter's magnetic field.

On the same day the atmospheric probe sampled Jupiter's atmosphere, the Galileo spacecraft began to orbit the planet, and went on to complete 35 elliptical orbits during eight years of observation. Each orbit was carefully planned to coincide with a flyby of Jupiter's moons. Galileo's discoveries in the Jovian system are numerous, including the presence of ammonia ice crystals in Jupiter's clouds; volcanic activity on Io that is 100 times greater than on Earth (making it the most geologically active object of the solar system); evidence of an ocean under Europa's surface; Ganymede's magnetic field; the origin of Jupiter's rings; and the form and changes of Jupiter's magnetosphere.

On September 21, 2003, after 22 years of good and loyal service, the Galileo mission was abruptly terminated when the orbiter was sent burning in Jupiter's atmosphere at a speed of 48.2 kilometers per second (equivalent to traveling the distance between Los Angeles and New York in 82 seconds). This was done primarily to avoid contamination of Jovian moons with bacteria, specifically Europa. Indeed, data from Galileo revealed that Europa might harbor life forms in the salt-water ocean under the its frozen surface.

GALILEO

(1564-1642)

Galileo Galilei, known simply as Galileo, was an Italian mathematician, physicist and astronomer born in Pisa. He is considered the true founder of the scientific method and the father of modern physics. He invented the thermometer and the hydrostatic balance to measure density, and he established the first laws of mechanics and ballistics. In 1609, he perfected the telescope built the year before by the Dutch optician Hans Lippershey (1570–1619). The quality of his rigorous astronomical observations led to several important discoveries, including mountains on the Moon, Jupiter's satellites (the Galilean moons), Saturn's ring, the phases of Venus, sunspots, the Sun's rotation and star clusters.

He is famous for defending the Copernican heliocentric model of the solar system against the geocentric idea defended by the Roman Catholic Church at that time. Despite being a fervent Catholic, at the age of 70 he had to appear in front of the Inquisition's tribunal and was forced to deny his findings. Legend has it that after being compelled to kneel down, he rose and stomped his foot on the ground declaring, "Eppur si muove!" ("and yet, it moves!").

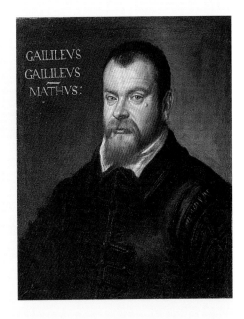

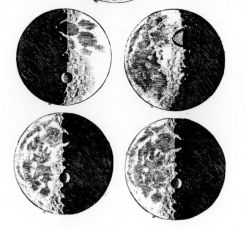

Cassini-Huygens

I N 1982, A YEAR AFTER the Voyager 2 flyby, planning began for an ambitious robotic mission to Saturn that would combine an orbiter and a descent module. The idea arose during a collaborative discussion between the European Science Foundation and the American Science Academy. In 1988, it was decided that NASA's Jet Propulsion Laboratory would build the Cassini orbiter (named after the Franco-Italian astronomer: see page 253), while the European Space Agency Technical Centre would construct the descent module, Huygens (named after the Dutch physicist and mathematician: see page 253).

RIGHT
The Cassini-Huygens probe is carefully lowered so that it can be attached to the adapter module of the Centaur rocket stage.

FAR RIGHT
Long-exposure photograph of the morning launch of the Cassini-Huygens mission by a Titan IV-B/Centaur rocket, on October 15, 1997, at the onset of a seven-year voyage to Saturn.

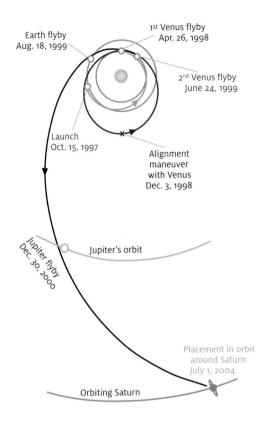

Earth flyby
Aug. 18, 1999

1st Venus flyby
Apr. 26, 1998

2nd Venus flyby
June 24, 1999

Launch
Oct. 15, 1997

Alignment
maneuver
with Venus
Dec. 3, 1998

Jupiter flyby
Dec. 30, 2000

Jupiter's orbit

Placement in orbit
around Saturn
July 1, 2004

Orbiting Saturn

Artist's interpretation of Cassini-Huygens approaching the Saturn system.

LEFT

The impressive size of the Cassini-Huygens spacecraft required that it take an indirect trajectory with considerable gravitational assistance. Two flybys of Venus, one Earth flyby and a passage close to Jupiter propelled it toward Saturn.

MAIN COMPONENTS OF THE CASSINI

1. Low-gain antenna
2. High-gain antenna
3. Radar module
4. Moveable platform to detect fields and cosmic dust
5. Magnetosphere imager
6. Huygens probe
7. Attitude-control thruster
8. Nuclear thermoelectric generator
9. Infrared spectrometer
10. Remote-sensing platform
11. Magnetometer boom (partial)

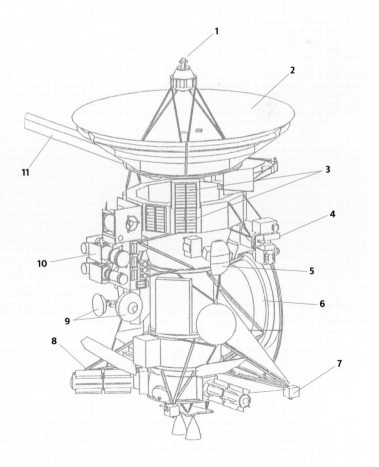

This most beautiful "portrait" of Jupiter was taken by the Cassini spacecraft during its brief visit en route to Saturn.
The mosaic is a composite of 27 telephoto images.

Despite the high cost of this project, the fear of negative political fallout if the collaboration broke down saved the mission on two occasions: in 1992 and in 1994, during American Congressional budget cuts. Hundreds of engineers and scientists from 16 European countries and 33 American states participated in the conception, construction, navigation and data analyses of this mission which continues to this day. It has been estimated that the mission has cost upwards to almost US$3.3 billion, which includes $1.4 billion in development, $755 million in operational costs, $54 million in tracking, $422 million for the launch vehicle, plus $500 million from ESA and $160 million from the Italian Space Agency.

The mission's main goal during several years of orbiting the giant planet was to determine the structure and composition of some key elements of the Saturnian system: in particular the rings, the surfaces of several satellites and Titan's atmosphere.

Saturn in all its cosmic glory!

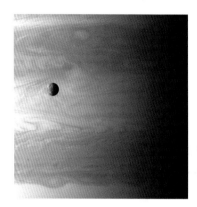

This image showing a transit of Io in front of Jupiter gives a misleading perspective, since the moon is actually 350,000 kilometers away from Jupiter, more than twice the planet's diameter.

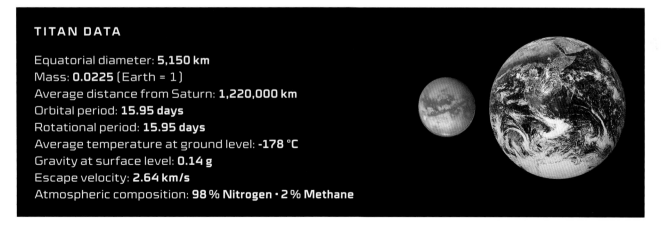

TITAN DATA

Equatorial diameter: **5,150 km**
Mass: **0.0225 (Earth = 1)**
Average distance from Saturn: **1,220,000 km**
Orbital period: **15.95 days**
Rotational period: **15.95 days**
Average temperature at ground level: **-178 °C**
Gravity at surface level: **0.14 g**
Escape velocity: **2.64 km/s**
Atmospheric composition: **98 % Nitrogen · 2 % Methane**

The purpose of this image is to capture the rings in their totality. The exposure time was adjusted to capture details in the shadowed areas, resulting in overexposure of Saturn.

Cassini's 3D attitude-control system uses three gyroscopes and an optical navigator with 5,000 reference stars. Propulsion for controlling the orientation relies on 16 hydrazine thrusters and 3,132 kilograms of fuel. The orbiter is equipped with a battery of scientific instruments: several optical spectrometers operating in different spectral bands, a mass spectrometer, radar to pierce Titan's opaque atmosphere, a magnetometer, a magnetosphere imaging system, a cosmic-dust analyzer and a plasma sensor. A radio transmitter, which is very stable both in frequency and strength, makes it possible to detect the minor changes in the probe's speed that are caused by gravitational field variations. The images are recorded on two one-megapixel CCD cameras, one equipped with a 200 mm f/3.5 wide-angle lens and the other with a 2,000 mm f/10.5 telephoto lens. Filters permit images to be recorded at different wavelengths. The on-board computer is based on an IBM 1750A microprocessor.

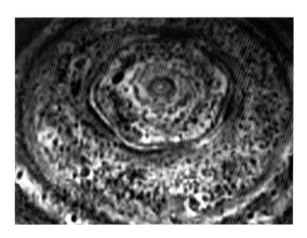

A hexagonal structure is located at Saturn's North Pole. This is a particularly intriguing, stable atmospheric phenomenon, which the Voyager probes had observed during their flybys nearly two decades earlier.

Details of Saturn's rings with enhanced colors to highlight the different densities (blue = denser, red = less dense). The red line in the external part of the ring outlines Encke's Division, named in honor of the 19th century German astronomer, Johann Encke.

LEFT

This striking panoramic view of the Saturn system totally eclipses the Sun, something that is possible to view from Earth. Protected from the Sun's rays by Saturn during this passage, Cassini was able to discover two new low-density rings.

A model of the Huygens probe in its final configuration on Titan's surface.

Securing the heat shield on the Huygens descent module.

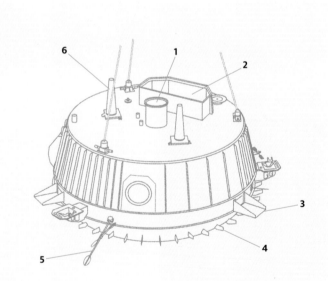

MAIN COMPONENTS OF HUYGENS

1. Ejection device (empty)
2. Parachute compartment (empty)
3. Radar altimeter antenna
4. Stabilizing fins
5. HASI atmospheric analyzer
6. S-band antenna

Huygens' parachute braking system was tested on a snowy plain in Sweden.

Two "solid state" recorders (without moveable parts), each with a 2.5-gigabyte capacity, hold the scientific data collected before transmission to Earth. The Huygens probe uses the S band for radio communications, the Ka band for radar signals, and the 8.4 Ghz X band, amplified to 20 watts, for communications with Earth. The Cassini orbiter is equipped with two low-gain antennas and one high-gain antenna, which is four meters in diameter. Electricity is provided by three nuclear thermoelectric generators (see page 248).

The Huygens probe, 2.7 meters in diameter, was designed to penetrate Titan's atmosphere and land on its surface. It consisted of an entry module and a descent module. The descent module, measuring 1.30 meters in diameter, carried six instruments. HASI (Huygens Atmosphere Structure Instrument) measured the physical properties (temperature, wind speeds, density) of Titan's atmosphere. DWE (Doppler Wind Experiment) measured wind speed as a Doppler effect, thanks to a very stable radio transmitter. The DISR (Descent Imager/Spectral Radiometer) was made up of an assembly of spectrometers, photometers and three digital cameras sharing the same CCD, to image Titan's surface and thermal emissions between the surface and the atmosphere. The GC/MS (Gas Chromatograph/Mass Spectrometer), a gas chromatograph coupled to a mass spectrometer, was used to analyze the atmosphere's chemical composition during the descent. The ACP (Aerosol Collector and Pyrolyser) captured droplets of aerosol and analyzed their chemical composition with the GC/MS. The SSP (Surface Science Package) used several sensors to characterize the physical properties (either liquid or solid) at the impact site on Titan's surface.

ABOVE, TOP
Artist's conception of the various descent phases of Huygens on Titan.

ABOVE
Aerial view of Titan, transmitted by Huygens during its descent.

The first photo of the surface of Titan. Huygens became the first probe to successfully land on an object in the outer solar system on January 14, 2005. For comparison, a same-scale image of the Moon's surface (right) gives an idea of the size of the rocks in the foreground.

This radar image of Titan by the Cassini orbiter shows a landscape of liquid methane lakes with very jagged shorelines.

At 7 meters in height, the Cassini-Huygens spacecraft is the most complex and imposing one ever launched by NASA. The orbiter and the Huygens module weigh 2,157 and 373 kilograms, respectively. Adding the launch adapter and fuel to that, the spacecraft weighed a total of 5,600 kilograms. In order to reach the required velocity, Cassini's trajectory required several gravitational boosts from Venus, the Earth and finally Jupiter, in what was termed the VVEJGA (Venus-Venus-Earth-Jupiter Gravitational Assist) trajectory. The Cassini-Huygens mission was launched on October 15, 1997 by a Titan IV-B/Centaur rocket from Cape Canaveral Air Force Station in Florida. Its first flyby of Venus took place on April 26, 1998, at a velocity of 11.7 kilometers per second and a mere 284 kilometers above the planet's surface. Its second flyby of Venus occurred on June 24, 1999, at 13.6 kilometers per second and an altitude of 600 kilometers. The spacecraft flew by Earth after that on August 18, 1999, at 19.1 kilometers per second and an altitude of 1,171 kilometers. On December 1, 1999, while on its way to Jupiter, the probe's high-gain antenna was pointed toward Earth.

A thin external layer appears to be floating above Titan's atmosphere. An ultraviolet filter is required to show it clearly.

Observations of Jupiter began on October 1, 2000. Two months later, the spacecraft received a gravitational assist while flying past Jupiter at a distance of 9.7 million kilometers and a velocity of 11.6 kilometers per second, thereby shortening the trip to Saturn by two years. Cassini took over 26,000 photographs of the Jovian system during this flyby, including some of the most detailed ever transmitted.

Engineers uncovered a critical error in the mission's communications system. Cassini's radio receiver had not been designed to capture radio signals from the Huygens probe when it was subjected to the Doppler effect, as would be the case during its descent onto Titan. That meant that much of the descent data obtained by Huygens during the descent stage would have been lost. It took engineers several more months to arrive at a satisfying solution to the problem. Cassini's trajectory was modified to minimize the Doppler effect during the data-relay stage and Huygens' software was reprogrammed to adjust to the new situation.

Cassini began its Saturn observations on February 6, 2004, and the following July, at the end of what was now a seven-year, 3.5-billion kilometer journey, it began to orbit Saturn and made its first crossing of the rings, through the gap separating the F and G rings.

On December 25, 2004, the Huygens probe was jettisoned via a spring-loaded mechanism, to begin a three-week trek directly toward Titan.

A Controversial Launch

The launch of the Cassini-Huygens mission was the subject of much controversy because it carried a thermoelectric nuclear generator fueled by plutonium-238 dioxide, a highly toxic radioactive element. Since this was the largest quantity of plutonium ever carried aboard a spacecraft, many citizens and physicists opposed the launch because they believed it carried a high potential risk of atmospheric dispersal in case of a catastrophic accident. It must be emphasized that the nuclear generators are designed to withstand such a launch explosion, since the nuclear pellets are encased in impact-resistant glass and the combustible marbles are themselves encased in protective graphite compartments. Officially, the probability of contamination was estimated at 1 in 1,400 during the first three minutes following ignition of rocket engines, at 1 in 476 during the rocket's ascent into the high atmosphere and at 1 in 1,000,000 in the case of a mishap during the planned Earth flyby two years later. Since plutonium-238 has a half-life of 87.8 years, opinions differed as to the number of potential cancer deaths from contamination in case of an accident. NASA had to make a very strong case to convince the general public that the risk of this mission was minimal. Both the launch and the flyby went off without any incident.

The full disk of Saturn's moon Iapetus was first photographed by the Cassini orbiter. The dark materials that covers the satellite's other side come from a large dust ring orbiting Saturn that was discovered in 2009. The large impact crater on the bottom left of the globe is 450 kilometers in diameter.

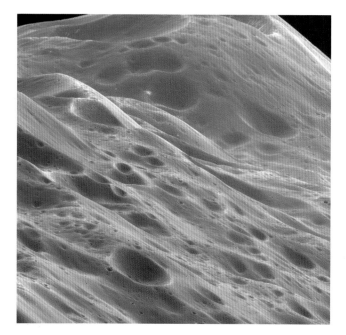

The equatorial mountains on Iapetus, Saturn's two-faced satellite, reach heights of 10 kilometers.

The strange, sponge-like appearance of Saturn's small satellite, Hyperion. The smallest visible element in this picture measures 362 meters.

OPPOSITE PAGE
The moon Enceladus is a world of ice, with relatively recent surface fractures (striations). Insert: Cassini observed geysers on Enceladus, spewing a mixture of water vapor, liquid methane and organic compounds into space.

On January 14, 2005, Huygens entered Titan's atmosphere. Four minutes later, as the probe was traveling at a speed of 1,440 kilometers per hour, its pilot parachute was deployed. After the latter was jettisoned, the main 8.5-meter diameter parachute was deployed. Once the heat shield was jettisoned, several instruments were activated, including the cameras. All Huygens data were transmitted to the Cassini orbiter, which relayed them to Earth. Another design error, this time in the communications program between Cassini and Huygens, resulted in the loss of one of the two Huygens radio channels. This cut in half (from 700 to 350) the number of images transmitted by Huygens to Cassini, as well as data on wind speeds. Approximately 14 minutes later, when the probe was about 160 kilometers from ground, the main parachute was jettisoned and a smaller, 3-meter diameter parachute took over so as not to slow the descent too much. The probe began to rotate on itself, providing a panoramic sweep for its cameras. At an altitude of 60 kilometers, Huygens' radar altimeters were activated to verify the probe's altitude. As it approached the surface, its on-board light came on to compensate for the low levels of luminosity. After a descent of 2 hours and 30 minutes, at the moderate speed of 5 to 6 meters per second, Huygens touched down intact in Titan's Adiri region.

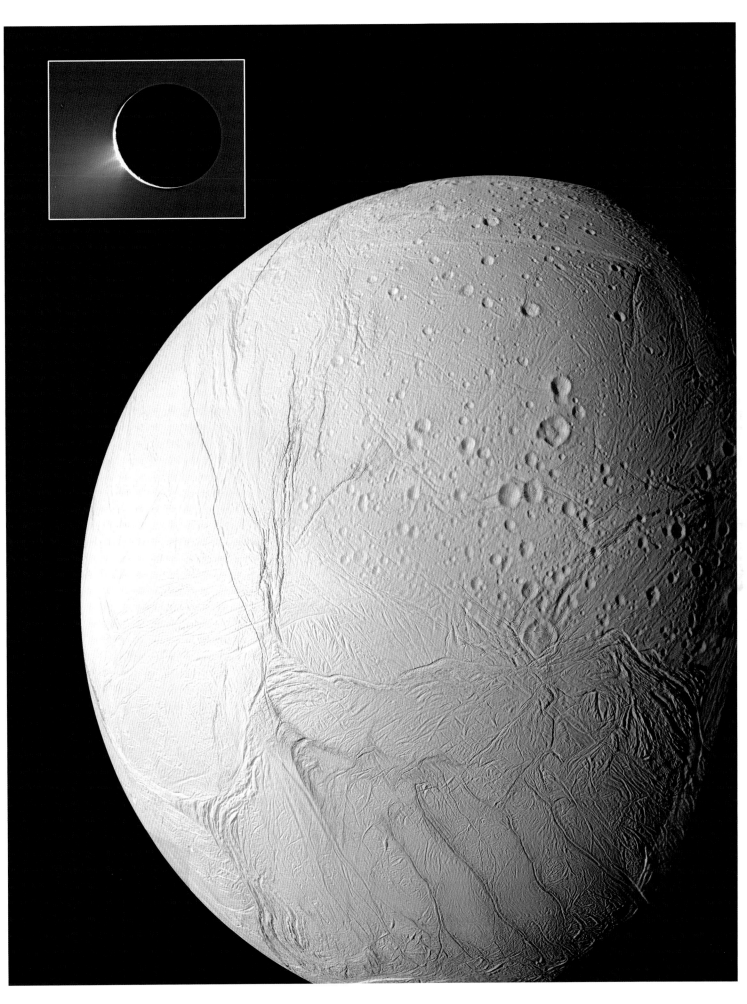

The icy and pale moon Dione, imaged while Cassini was approaching the plane of the rings whose striated shadow is projected on Saturn.

The surface had the consistency of corrugated ice. The lander was able to transmit one image from this location and various data for 90 minutes after landing.

For its part, Cassini became the first spacecraft to orbit Saturn. From a distance of 10 AU, radio signals take 84 minutes to travel from Earth to Cassini and vice versa. For that reason, engineers in charge of the probe's navigation cannot control Cassini in real time, and must wait more than three hours before receiving a response from the probe for both routine day-to-day operations and for urgent matters. Despite such technical challenges, the scientific data obtained from Cassini are abundant: accurate determination of Saturn's rotation period, elucidation of the composition of the rings, discovery of the eye of a cyclone in Saturn's South Pole and discovery of three new moons. Cassini has made the first flyby of the satellite Phoebe as well as multiple flybys of Enceladus. And, during several flybys, Cassini established the first topographic map of Titan, using its on-board radar. Many structures oddly reminiscent of terrestrial lakes and rivers were also observed.

NASA has announced an extension of this remarkable mission, till 2017. NASA does not plan to crash Cassini into Saturn to destroy it at the end of the mission, the risk being too high that Cassini might strike an object while crossing the rings and become uncontrollable. Scientists are working diligently on a solution to avoid contaminating Titan and Enceladus with radioactive debris. One option is to place the probe in a high stationary orbit and then to have it crash into a small moon where the plutonium contamination would be inconsequential.

GIOVANNI DOMENICO CASSINI

(1625–1712)

At the behest of French finance minister, Jean-Baptiste Colbert, the Italian astronomer Cassini moved to France in 1669, where he became the Director of the Paris Observatory in 1671. He was the first to describe Jupiter's Great

Red Spot in 1665 and to calculate the planet's speed of rotation. In addition, he discovered the discontinuity between Saturn's A and B rings (the Cassini Division) as well as four of the planet's satellites. In 1673, during a transit of Venus across the Sun, Cassini accurately determined the distance from Earth to the Sun. His accurate descriptions of the Moon, Mars, Jupiter and Saturn are among the classic discoveries in astronomy. He was responsible for a dynasty of astronomers: his son Jacques (Cassini II, 1677–1756) and his grandson César François (Cassini III, 1714–1784) followed in his footsteps as the directors of the Paris Observatory.

CHRISTIAAN HUYGENS

(1629–1695)

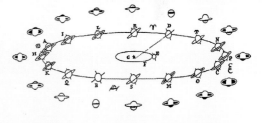

Born in The Hague, Huygens was a physicist, clockmaker, mathematician and astronomer. In physics, he developed an interest in oscillating pendulums as instruments to measure time and weight. He also invented and built pendulum clocks with an ingenious escape mechanism, still in use to this day. He was among the first to postulate the wave theory of light. In the area of mathematics, encouraged by Blaise Pascal, he published the first complete treatise of probability calculations in 1657. Among his contributions to astronomy are the improvement of telescopes through better eyepieces and the discovery of the rings of Saturn in 1655, as well as its giant moon, Titan. He also determined Mars' rotation period and observed the many stars in the Orion nebula.

New Horizons

IN ORDER TO BE ABLE TO VISIT PLUTO in 2015, the New Horizons Probe (see page 335) followed a trajectory that brought it close to Jupiter, where it would benefit from a gravitational boost. That flyby took place on February 28, 2007, at a speed of 23 kilometers per second. Jupiter's gravitational assist added 4 kilometers per second to the probe's velocity, thereby shortening the Earth to Pluto voyage by two years. New Horizons' digital telescope LORRI (Long Range Reconnaissance Imager) was tested on the Jovian system in September 2006 from a distance of 300 million kilometers, with spectacular results! As the spacecraft flew by Jupiter at 3 million kilometers, it studied the planet's cloud dynamics and provide close-up observation of the storm Oval BA, a southern hemisphere feature first observed in 2000 and nicknamed the "Small Red Spot" after it took on a reddish coloration in August of 2005. Although the Galilean moons were not in favorable positions during New Horizons' brief passage, its instruments were calibrated for low luminosity targets and its cameras, of higher quality than those on the Galileo probe, were able to produce remarkable images and spectrometric data of Io, Ganymede and Europa.

This composite of several images, taken with the New Horizons multi-spectral camera, illustrates the remarkable diversity of the atmospheric features on Jupiter.

RIGHT
Artist's rendition of New Horizons during its journey, with Io and Jupiter in the background.

OPPOSITE PAGE
Photomontage of two images of Jupiter and the volcanic moon Io, as transmitted by New Horizons during its February 2007 flyby.

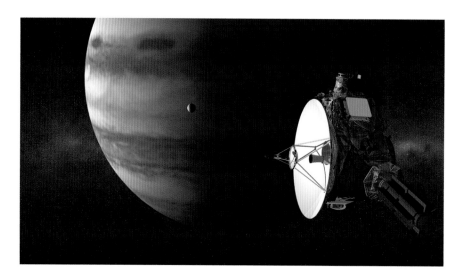

Mercury ☿

Since Pluto has been classified as a dwarf planet, Mercury has now risen to the rank of the smallest planet and the one closest to the Sun. Because of its proximity to the Sun, this dim planet can only be seen from Earth at dawn and at dusk. Sumerians were the first to describe Mercury, 3,000 years BCE. Assyrian writings dating to 14 centuries BCE described it as "the planet with odd movements." In ancient Greece, astronomers thought they were seeing two distinct stars, Apollo at dawn and Hermes at sunset, before realizing, four centuries BCE, that they were indeed viewing a single object. Because this planet moves so quickly in the sky, the Romans assigned it the name of Mercury, the equivalent of the Greek god Hermes, in honor of the messenger of the gods. The name Mercury is also given to Wednesday, the third day of the week, in several languages: *mercredi* in French, *mercoledi* in Italian, *miércoles* in Spanish and *Mercurii dies* in Latin.

Mercury remains a mysterious planet. The Messenger mission was the first to attempt its complete mapping.

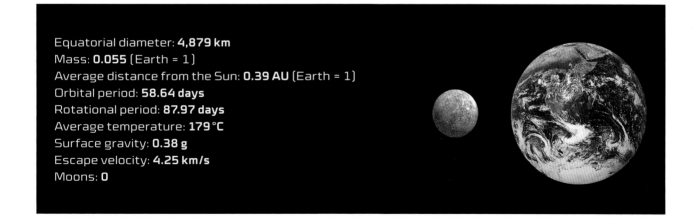

Equatorial diameter: **4,879 km**
Mass: **0.055** (Earth = 1)
Average distance from the Sun: **0.39 AU** (Earth = 1)
Orbital period: **58.64 days**
Rotational period: **87.97 days**
Average temperature: **179 °C**
Surface gravity: **0.38 g**
Escape velocity: **4.25 km/s**
Moons: **0**

Mercury turns three times on its axis over the course of two orbits around the Sun, therefore three Mercurian "days" last two Mercurian "years." If an observer were standing on this planet, he would experience 176 Earth days between two sunrises.

Though quite different internally, Mercury resembles our Moon in external appearance, with its pockmarked surface of craters and lack of atmosphere. A metallic core makes up more than 40 percent of its volume, which accounts for Mercury's high density and its magnetic field. Mercury's orbit is a highly eccentric ellipse, with a radius that varies from 46 to 76 million kilometers. Because of that, it took astronomers a long time to ascertain Mercury's orbit, and the gradual rotation of the ellipse (perihelic precession) remained a mystery for quite some time. Some astronomers even went so far as to postulate the existence of a disruptive planet named Vulcan. Finally, in 1925, a then little-known German physicist, Albert Einstein, provided an exact calculation of Mercury's orbit and its precession. The solution of this enigma was one of the first convincing proofs of the general theory of relativity; something widely reported in the media at that time, which indirectly elicited interest in Einstein's other theories.

Detailed telescopic observations of Mercury continue to be challenging, and only two space probes have been able to visit the planet so far. Mariner 10 flew by in 1974 and 1975 and Messenger came close in 2008 and started to orbit the planet in March 2011. All data transmitted by robotic probes in the years to come will be crucial to our knowledge about this as yet poorly understood planet. ☿

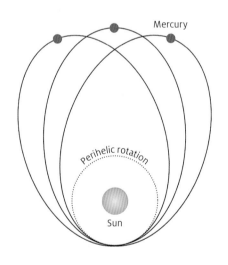

Without using a telescope, Albert Einstein (1879–1955) applied his general theory of relativity to explain the slow rotation of Mercury's perihelion, which had long been observed (but not understood) by astronomers.

Mariner 10

THE FIRST SPACE-PROBE FLYBY of Mercury occurred on March 29, 1974, when Mariner 10 approached the planet to within 700 kilometers, following a visit to Venus a few weeks earlier (see page 101). After a long trip around the Sun, Mariner 10 again passed close by Mercury, this time to within 48,000 kilometers, on September 21, 1974. Since the spacecraft's attitude-control system gyroscopes had failed, its moveable solar panels took over the role of solar sails in order to control the probe's attitude without burning fuel. The third and last encounter, some 327 kilometers above Mercury, put an end to this mission on March 16, 1975. The data transmitted by Mariner 10 completely transformed our image of Mercury, which has always been difficult to observe with an Earth-based telescope. The fact that Mariner 10's orbital period was exactly twice that of Mercury meant that the probe observed the same illuminated side of the planet on three separate occasions and it was only able to photograph and map 40 to 45 percent of the planet's surface. The 3,400 images obtained revealed a lunar-like surface studded with craters. Mariner 10 also discovered a tenuous helium atmosphere on Mercury, a weak magnetic field and an iron-rich core. It measured a maximum diurnal temperature of 427°C and a nocturnal temperature of −183°C. Mariner's last flyby was in 1975, and Mercury had to wait until 2008 for the next visit, this time by the Messenger probe (see page 262). ☿

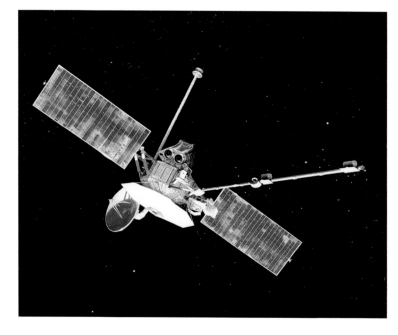

Mariner 10 was the first space probe in history to visit the planet Mercury.

A composite of 18 photographs taken while Mariner 10 was approaching Mercury for the first time, on March 29, 1974. Mercury's surface, pockmarked with craters, is strangely reminiscent of the Moon's.

Reconstruction of Mercury's globe, based on data obtained from the Mariner 10 mission.

LEFT

The metallic core of Mercury occupies a large portion of its internal volume, and this explains the strong density of this small, telluric planet.

2004

Messenger

One of the final preparatory steps in readying the Messenger probe: filling the fuel tanks before encapsulation into the rocket's nose cone.

MESSENGER IS THE FIRST MISSION to visit Mercury in more than 30 years. The acronym "Messenger" stands for MErcury Surface, Space ENvironment, GEochemistry and Ranging mission, as well referring to the fact that Mercury is the messenger of the gods in Roman mythology. The goal of the mission was to map the entire surface of the planet, to characterize its surface composition and thin atmosphere, and to gather enough information to be able to understand the planet's geologic past and the origins of its magnetic field.

The Delta II rocket carrying the Messenger lifted off on August 3, 2004, from Cape Canaveral. Initially scheduled for takeoff in May of 2004, NASA was forced to delay the mission until August due to adverse weather conditions. This change in plans compelled them to make a complete trajectory revision, which resulted in a two-year delay for Messenger's arrival. Mercury is by no means an easy destination, due to its proximity to the strong gravitational influence of the Sun. A spacecraft is accelerated in the Sun's direction and then must be slowed down in order to reach Mercury. Because of this, Messenger's trajectory required considerable gravitational assistance. Three flybys were needed initially: an Earth flyby (August 2, 2005) and two flybys of Venus (October 24, 2006 and June 5, 2007), including one at only 332 kilometers from the surface. The first flyby of Mercury occurred on January 14, 2008, followed by a second one on October 6, 2008. The last Mercury flyby, on September 29, 2009 further slowed the probe. Since Mercury does not have enough of an atmosphere to permit aerobraking, Messenger had to rely entirely on its engines to slow down enough for capture into orbit in March 2011.

The space probe was designed and assembled at the Applied Physics Laboratory of John Hopkins University in Baltimore. It is equipped with six observation and measuring devices.

Artistic interpretation of the
Messenger probe orbiting Mercury.

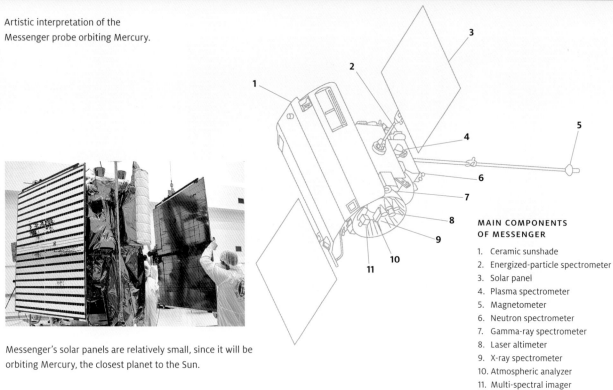

Messenger's solar panels are relatively small, since it will be
orbiting Mercury, the closest planet to the Sun.

MAIN COMPONENTS OF MESSENGER

1. Ceramic sunshade
2. Energized-particle spectrometer
3. Solar panel
4. Plasma spectrometer
5. Magnetometer
6. Neutron spectrometer
7. Gamma-ray spectrometer
8. Laser altimeter
9. X-ray spectrometer
10. Atmospheric analyzer
11. Multi-spectral imager

Messenger took advantage of an Earth flyby in August 2005, during a gravitational assist maneuver, to successfully test its imaging system.

The MDIS (Mercury Dual Imaging System) consists of two digital cameras, one of which is coupled to a wide-angle lens, and the other to a telephoto lens for close-up images of the surface. It is anticipated that this will provide precise stereoscopic image coverage of the entire planet, at a resolution down to 250 meters per pixel. The GRNS (Gamma Ray and Neutron Spectrometer), the MASCS (Mercury Atmospheric and Surface Composition Spectrometer) and the XRS (X-Ray Spectrometer) will be studying the chemical composition of Mercury's soil and atmosphere. The EPPS (Energetic Particle and Plasma Spectrometer) measures the properties of the charged particles in the magnetosphere. The MLA (Mercury Laser Altimeter) will contribute to surface relief mapping. A magnetometer, located at the end of a 3.6-meter boom, will measure magnetic fields. Its attitude control system consists of an assembly of gyroscopes and accelerometers, optical navigation trackers and propellant thrusters. Radio communication with Earth occurs in the X band. A ceramic heat shield serves as a solar screen to maintain temperatures within operational range when the spacecraft is exposed to the Sun. Two rotatable solar panels can supply 450 watts of power, and several radiators in the body of the probe regulate the thermal flow that could interfere with measurements of the infrared radiation emitted by Mercury. Messenger is also equipped with two control computers to provide redundancy in case one fails, each outfitted with radiation-resistant IBM RAD6000 processors.

A series of 614 photographs, taken during the approach and flyby of Venus (July 5, 2007), were used to calibrate cameras in preparation for the Mercury-centered mission.

This unique impact crater with several radiating faults (nicknamed the Spider), lies at the center of Mercury's Caloris basin, and remains perplexing to scientists.

Volcanic activity has played a critical role in remodeling Mercury's surface, as illustrated in these lava-filled craters.

Similar to the PowerPC processors in old Macintosh Apple computers, they are rated at 25 MHz. For data storage, Messenger has two solid-state recorders, each with one gigabyte capacity. The RAD6000 processors will collect, compress and store images and other data obtained by Messenger's instruments before their transmission to Earth.

The three Mercury flybys (January 2008, October 2008 and September 2009), at altitudes between 200 and 230 kilometers, have already made several impressive discoveries. These include proof of past volcanic activity on Mercury, observations of radiating impact craters, evidence of a shifting liquid core that generates the magnetic field and a significant amount of water vapor in its exosphere (very high atmosphere). Messenger has already imaged 30 percent more of the surface than previously observed by Mariner 10. Between the Mariner 10 mission and the three passages by Messenger, more than 90 percent of the planet's surface has been photographed. Successfully in orbit since March 18, 2011, Messenger will observe Mercury for at least another year, which should help clarify more of the planet's mysteries. ☿

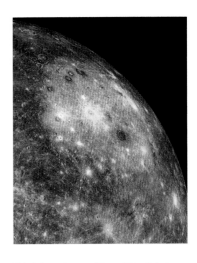

This false-color rendition of the Caloris Basin, also known as Caloris Planitia, differentiates its metallic components from the surrounding plains. The orange spots, located at the periphery of the 1,550-kilometer basin, denote volcanic structures.

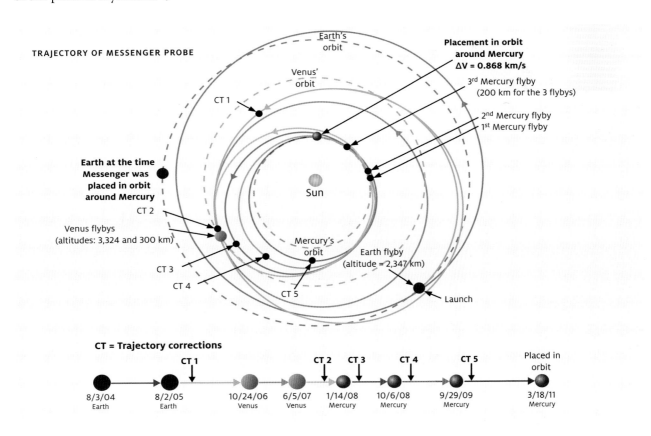

TRAJECTORY OF MESSENGER PROBE

Earth's orbit

Venus' orbit

CT 1

Placement in orbit around Mercury
ΔV = 0.868 km/s

3rd Mercury flyby (200 km for the 3 flybys)

2nd Mercury flyby
1st Mercury flyby

Earth at the time Messenger was placed in orbit around Mercury

CT 2

Venus flybys (altitudes: 3,324 and 300 km)

CT 3

CT 4

CT 5

Sun

Mercury's orbit

Earth flyby (altitude = 2,347 km)

Launch

CT = Trajectory corrections

| CT 1 | | | | CT 2 | CT 3 | | CT 4 | | CT 5 | Placed in orbit |

| 8/3/04 Earth | 8/2/05 Earth | 10/24/06 Venus | 6/5/07 Venus | 1/14/08 Mercury | 10/6/08 Mercury | 9/29/09 Mercury | 3/18/11 Mercury |

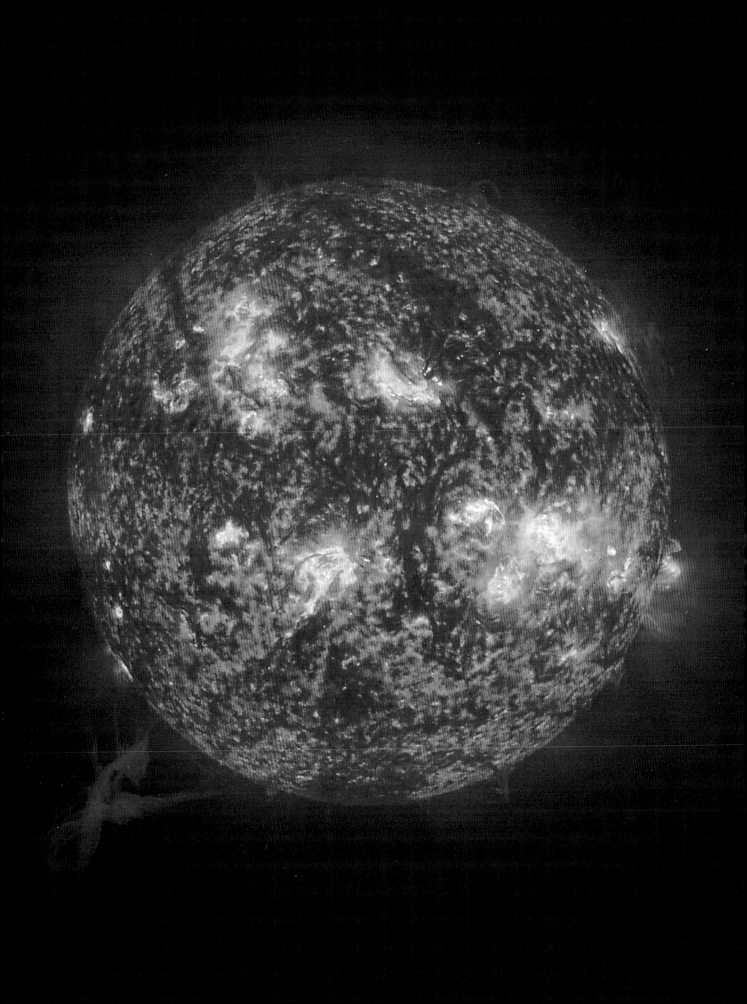

The Sun ◉

CLASSIFIED AS A YELLOW DWARF STAR, the Sun lies at the center of our solar system. The energy emitted by the Sun in the form of both light and heat supports life on Earth and determines climate and meteorological phenomena. The distance from the Sun to the Earth is 150 million kilometers, and it takes 8.3 minutes for the Sun's light to reach us on Earth.

All planets, asteroids and comets orbit the Sun, whose mass accounts for 99 percent of the entire mass of the solar system. The Sun itself is located some 25,000 light years from the center of our galaxy, the Milky Way, and orbits it at a velocity of 250 kilometers per second, completing one revolution every 250 million years. Of the 50 nearby stars, the Sun is the fourth in terms of luminosity and is classified as a medium-size star among the two hundred billion stars in the galaxy. The Sun appears yellow as seen from Earth because atmospheric absorption removes the blue component of sunlight. This is even more apparent when light must travel long distances through the atmosphere or when there is a significant amount of dust in the air. That's why the Sun appears orange-red at both sunrise and sunset. The Sun's spectral classification is G2V. G2 refers to the fact that its surface temperature is about 5,500°C and the V (5 in Roman numerals) indicates that it is a main sequence star that generates energy by converting its hydrogen to helium. These nuclear fusion reactions occur in its core, where the temperature approaches 15,000,000°C.

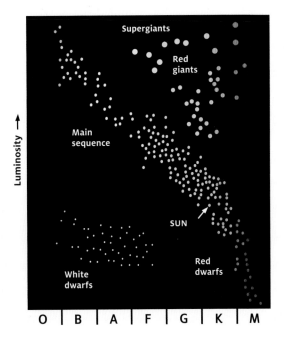

The Hertzsprung-Russell classification of stars, including the Sun, based on temperature and luminosity.

Like a thermonuclear bomb in a continuous state of explosion, the Sun burns more than four million tons of matter per second. Hydrogen makes up 74 percent of its mass and 92 percent of its volume, while helium represents only 24 percent of its mass and 7 percent of its volume.

At its present age of 4.6 billion years, the Sun has used up between 35 and 45 percent of its hydrogen reserves. At the anticipated end of its life cycle, in 5 billion years, the Sun will turn into a Giant Red, and at 160 times its present diameter, will engulf both the Earth and Mars. It will then eject its external layers into a planetary nebula and end its life as a small star of the white dwarf category. The high-energy photons generated through the fusion reactions in the Sun's core are slowed by the higher density of the upper layers, to the point that it takes them more than 10,000 years to escape in the form of sunlight! The Sun's visible light arises from the photosphere, a 400-kilometer-thick layer above the convective zone (see diagram). The solar atmosphere consists of the external layers of the photosphere, the chromosphere and the corona, all normally invisible (although they can be seen during an eclipse). The chromosphere is about 2,000 kilometers thick, with a temperature gradient that rises to 20,000°C from base to top. A thin layer, called the transition zone, where the temperature rises to 1,000,000°C, lies between the chromosphere and the corona. Forming the external layer of the solar atmosphere, the corona consists of tenuous plasma that mingles with the solar wind. It is in this zone that one finds extreme temperatures of 2,000,000°C.

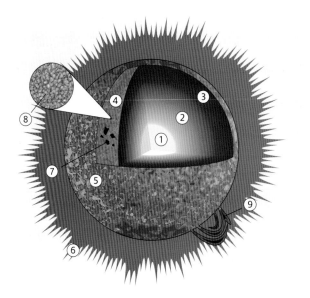

INTERNAL STRUCTURE OF THE SUN

1. Core
2. Radiative zone
3. Convective zone
4. Photosphere
5. Chromosphere
6. Corona
7. Sunspot
8. Granules
9. Prominence

It is through the corona that enormous bubbles of plasma are extruded; called coronal mass ejections (CME), they contain billions of tons of matter.

It is possible to observe the sunspots, which are darker than their surroundings, by using the appropriate filters. These spots correspond to colder regions with strong magnetic fields. The Sun rotates on its own axis, but since it is composed of gas and plasma, its rotation period varies between 25 days at the equator to 34 days in the polar regions. This torsion effect causes the Sun's magnetic field to assume a spiral shape, and over time, these magnetic field lines become twisted. This results in field-loop eruptions from the solar surface, which cross one another and trigger the formation of sunspots, prominences and dangerous solar eruptions. These magnetic phenomena follow an 11-year solar cycle that impacts the Earth's climate. The so-called Little Ice Age was a period of cold climate in Europe and North America between 1550 and 1860, that coincided with weakened solar activity.

A portion of the Sun's magnetic field is permanently dissipated in the form of solar wind. This flow of charged particles reaches phenomenal speeds in the order of a million kilometers per hour, traversing the distance between the Sun and Earth in a matter of days. When the solar wind encounters our magnetosphere, most of the charged plasma is deflected by the Earth's magnetosphere, but some of it follows magnetic field lines into the polar regions. These particles energize the gas molecules in the upper atmosphere, which emit photons and create the spectacular light shows known as aurora borealis (the northern lights) and aurora australis (the southern lights).

At the onset of the space age, the first robotic probes, Luna and Pioneer, were launched to explore the interplanetary space environment, which led to discovery of the solar wind. Several Earth-orbiting satellites (Solar Maximum Mission, Yohkoh, Viking, Integral, Odin, Trace, Cluster, Double Star) and the solar probes monitor the Sun so that we can learn more about our own star and can predict variations in its activity that affect our home planet. ⊙

This Pieter Bruegel the Elder painting, entitled Hunters in the Snow, dates back to 1565. At that time, Europe was undergoing a minor ice age, probably due to disruptions in solar activity.

Pioneer 5

PIONEER 5, the second American space probe, was a sphere-shaped craft weighing 43 kilograms and was launched on March 11, 1960, by a Thor-Able rocket. Its mission was to explore the interplanetary space between Earth and Venus. Data on magnetic fields, solar wind and cosmic radiation were transmitted by radio signals at a rate of either 1.8 or 64 bits per second, depending on the distance between the Earth and the probe and the size of the receiving antenna. More than 3 megabits of scientific data have been transmitted by Pioneer 5. Its last detectable signal was received on June 26, 1960, by the radio telescope at Jodrell Bank in England, from a distance of 36 million kilometers, a record at the time. ☉

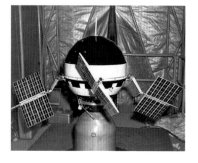

Pioneer 5 in its final configuration with its solar panels deployed. At right, several mission officials examine the probe, set on its Thor-Able rocket.

Pioneer 6 to 9

1965

O N DECEMBER 6, 1965, the small cylindrical probe known as Pioneer 6 was launched on a Delta-E rocket from Cape Canaveral and placed in a solar orbit at a distance of 0.8 AU. Its mission, to observe solar activity and detect cosmic rays, was to have lasted six months; in 2000, to commemorate its 35th birthday, Pioneer 6 was successfully contacted and it was still functioning, making it the oldest NASA spacecraft in operation! The subsequent launches of identical probes — Pioneer 7, 8 and 9 — between August 1966 and November 1968 (a fifth mission failed at launch in August 1969), created a vast space network for monitoring solar activity.

On December 16, 1965, a Delta-E rocket launched Pioneer 6, the first meteorological solar mission.

LEFT

A flotilla of small, 146-kilogram Pioneer probes was placed in a reconnaissance heliocentric orbit to monitor solar activity on a permanent basis, in order to alert space agencies in case of a magnetic storm. And this has lasted for more than 40 years!

The probes were spin-stabilized by rotating at 60 turns per minute, with their axes perpendicular to the plane of the ecliptic. The on-board instrumentation was specifically designed to provide the very first detailed description of the solar wind, the Sun's magnetic field and cosmic rays. The high-gain antenna provides radio communications at rates of 8 to 512 bits per second. Thanks to a few hours of lead time on Earth, Pioneer data provides advance warning in case of a solar storm, useful for Earth-orbit crewed missions and future lunar missions. ⊙

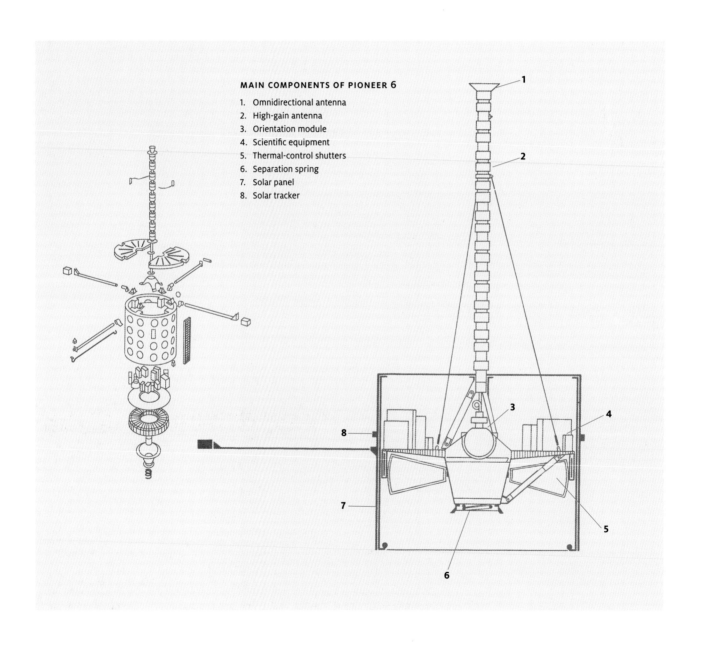

MAIN COMPONENTS OF PIONEER 6

1. Omnidirectional antenna
2. High-gain antenna
3. Orientation module
4. Scientific equipment
5. Thermal-control shutters
6. Separation spring
7. Solar panel
8. Solar tracker

Explorer 35

Nasa's s EXPLORER 35 PROBE was designed to study the Sun's effects on the properties of interplanetary space. It was part of the IMP (Interplanetary Monitoring Platform) scientific missions. On July, 19 1967, it was launched by a Thor-Delta rocket into an elliptical lunar orbit. By spin-stabilizing its axial rotation at 25.6 turns per minute, the spacecraft was oriented almost perpendicular to the plane of the ecliptic, allowing it to monitor interplanetary plasma, magnetic fields, micrometeorites and solar X-rays. This led to the discovery that the Moon creates a cavity in the solar wind. Another Explorer 35 experiment consisted of aiming radar waves at the Moon to measure the reflectivity of its surface. The reflected radar waves were detected on Earth via a 50-meter diameter antenna located in Palo Alto, California. The Explorer mission lasted six years and was deliberately terminated on June 24, 1973. ⊙

The 230-kilogram Explorer 35 spacecraft investigated the properties of interplanetary space from a lunar orbit.

Helios 1 and 2

THE SOLAR MISSION HELIOS was a collaboration between the Federal Republic of Germany (West Germany) and the United States. The Helios 1 and 2 (also known as Helios A and B) probes were built in Germany, while their launch was under NASA's direction. Helios 1 was launched on December 10, 1974, by a U.S. Air Force Titan III-E/Centaur rocket on a heliocentric elliptical orbit. Its mission was to measure the solar wind and explore the interplanetary space between the Earth and inside Mercury's orbit. The twin probe, Helios 2, followed on January 15, 1976, and entered a similar orbit. The scientific equipment carried by both Helios missions consisted of a magnetometer, a micrometeorite detector and a 32-meter-long dipole antenna to detect magnetic and electrical signals across a broad range of frequencies. Although the two probes' primary mission was completed at the beginning of the 1980s, they continued to transmit data until 1985.

The Helios probes set two significant historical records. First, they were the fastest human-made objects ever launched, with a velocity of 253,000 kilometers per hour (70 kilometers per second). Second, they went closer to the Sun than any other spacecraft at that time, reaching an estimated distance of 0.3 AU (45 million kilometers). ⊙

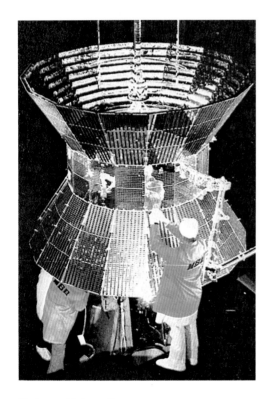

Final assembly inspection.

The Helios probe being encapsulated in the launcher nose cone.

The Titan/Centaur 2 launch vehicle with Helios on board, awaiting countdown at the U.S. Air Force Complex 41 launch pad on Cape Canaveral.

MAIN COMPONENTS OF THE HELIOS SPACECRAFT

1. Omnidirectional antenna
2. Medium-gain antenna
3. High-gain antenna and reflector
4. Solar panel
5. Gas reservoir
6. Attitude control
7. Search coil magnetometer
8. Single coil magnetometer
9. Micrometeorite analyzers
10. Photometer
11. Fluxgate magnetometer

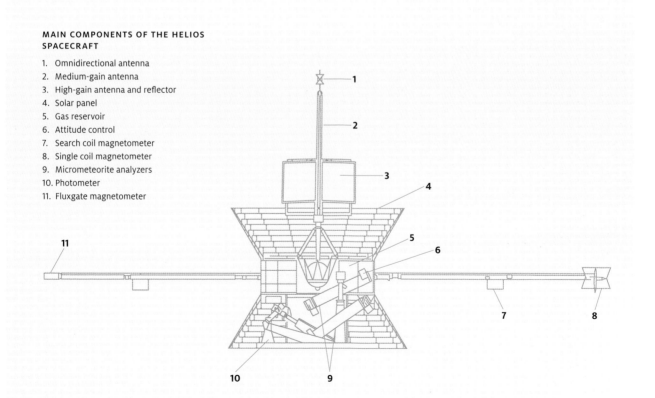

1978 ISEE-3/ICE

Artist's interpretation of ISEE-3/ICE.

THE ISEE (INTERNATIONAL SUN-EARTH EXPLORER) mission, an international collaboration between NASA and the European Space Agency (ESA), consisted of three spacecraft: ISEE-1 and ISEE-2 (a pair of satellites in geocentric orbits) and the space probe ISEE-3. ISEE-3 was launched on August 12, 1978, by a Delta rocket, and became the first probe in history to orbit at the Lagrangian point L_1 of the Earth–Sun system, which is located 1.5 million kilometers away from Earth in the direction of the Sun. The Lagrangian points are unique points in space where there is an equilibrium between the Earth's and the Sun's gravitational fields.

MAIN COMPONENTS OF ISEE-3/ICE

1. S-band antenna
2. Inertia stabilizer
3. 3D antenna (a total of 4)
4. Magnetometer
5. Solar panel
6. X-ray detector
7. Cosmic-ray detector

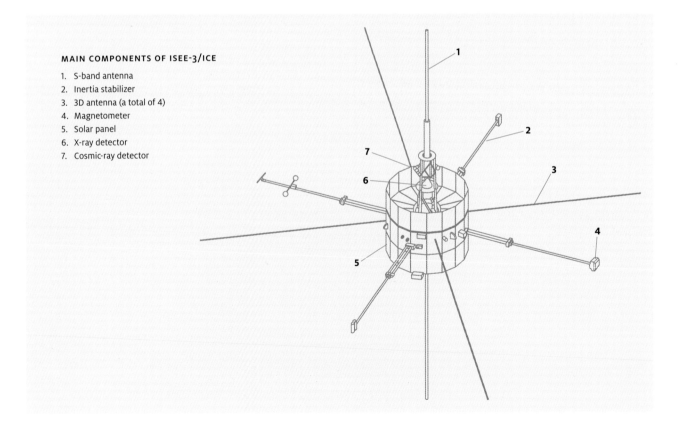

The ISEE-3 mission demonstrated the technical feasibility of reaching the stable orbit known as "halo" at the L_1 point. This served as an ideal site for solar observation for many years, from 1978 to 1982. The main goals of the mission were to study the solar wind in a region at the very limit of the Earth's magnetosphere and to accumulate data on cosmic rays and solar flares from a vantage point of about one AU. On June 10, 1982, ISEE-3 ended its first mission.

On September 1, 1982, it was reassigned a new mission, centered on the exploration of interactions between the solar wind and comet tails, and it was renamed ICE (International Cometary Explorer). After firing its engines to leave L_1 for a heliocentric trajectory toward comet Giacobini-Zinner, it passed through its tail on September 11, 1985. In March 1986, ICE passed between the Sun and Halley's Comet. At a distance of 28 million kilometers, it was quite far away compared to the "Halley Armada" (see the Giotto mission on page 301). In 1991, NASA redirected the ICE mission to monitor solar activity, specifically the detection of coronal flares. The mission was terminated in May of 1997, but then reactivated again in September 2008 by the Deep Space Network, after a detailed check of its instruments revealed that nearly all of them were in perfect working order. Two scenarios were then proposed: the first was to capture the probe during its close flyby of Earth in 2014 — but that was not possible since the space shuttle would no longer be available for this operation. The second option would involve sending the probe to comets in 2017–2018, thereby delaying its potential recovery until the 2040s. ⊙

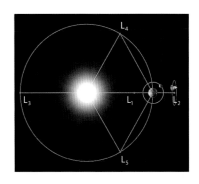

ISEE-3 was the first probe sent into orbit at a Lagrangian point (in this case L_1), which demonstrated the usefulness of these areas, where gravitational forces are in a state of equilibrium.

Mechanical inspection of ISEE-3 at NASA's Goddard Space Flight Center. These tests were conducted in a "white room" (environmentally clean room), so as not to affect the sophisticated instrumentation installed on the spacecraft.

Ulysses

On October 6, 1990, Ulysses was lifted into a low terrestrial orbit by the space shuttle Discovery.

T HE ROBOTIC MISSION ULYSSES was designed to study the Sun at all latitudes, particularly at the polar regions, which cannot be observed from Earth. The name Ulysses refers to the mythical voyager of Homer's *The Iliad* and *The Odyssey* and Dante's *Divine Comedy*, a hero who did not fear exploring new frontiers. The Ulysses mission was the fruit of collaboration between the European Space Agency (ESA), which developed the probe, and NASA, which was responsible for the launch and for providing the nuclear thermoelectric generator. Mixed teams based at NASA's Jet Propulsion Laboratory were in charge of managing the mission. Ulysses carried an impressive array of scientific instruments for a number of tasks: to measure the magnetic fields and the speed of the solar wind, to study the electrons, protons and ions emanating from the Sun, and to study cosmic rays originating in interstellar space.

The impetus and speed required to propel a probe directly above the Sun's poles is beyond the capacity of today's launch vehicles. Previously, gravitational-assisted maneuvers (such as those carried out by the Mariner 10 and Pioneer 11 probes, as well as by Voyager 1 and 2 during the 1970s) were limited to trajectory changes in the plane of the elliptic. For the Ulysses mission, a precisely calculated Jupiter flyby was proposed, to generate a slingshot effect and send the probe out of the plane of the ecliptic into a new trajectory above and below the Sun's poles.

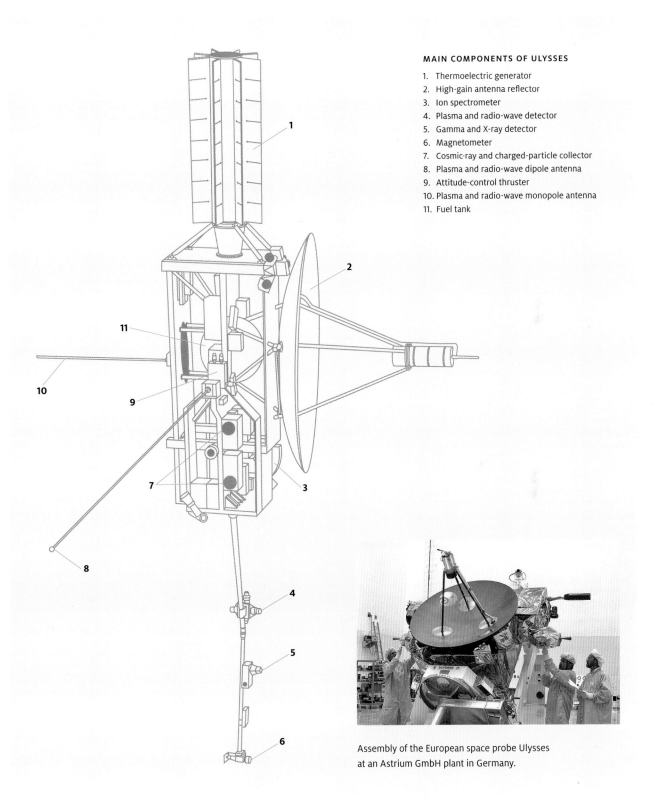

MAIN COMPONENTS OF ULYSSES

1. Thermoelectric generator
2. High-gain antenna reflector
3. Ion spectrometer
4. Plasma and radio-wave detector
5. Gamma and X-ray detector
6. Magnetometer
7. Cosmic-ray and charged-particle collector
8. Plasma and radio-wave dipole antenna
9. Attitude-control thruster
10. Plasma and radio-wave monopole antenna
11. Fuel tank

Assembly of the European space probe Ulysses
at an Astrium GmbH plant in Germany.

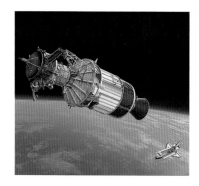

Artist's representation of the Earth-orbiting Ulysses on its booster rocket. The booster's two stages, IUS (Inertial Upper Stage) and PAM (Payload Assist Module), provided the necessary lift to place Ulysses in an Hohmann transfer orbit (see page 19) toward Jupiter.

The mission was scheduled to begin in 1983 but organizational delays postponed the scheduled launch to 1986, aboard the space shuttle Challenger. The loss of the Challenger in January of 1986 further delayed the mission to October 6, 1990, when Ulysses became part of the shuttle Discovery's payload on the STS-41 mission. After being jettisoned into low Earth orbit, Ulysses was propelled into a transfer trajectory directly toward Jupiter by a pair of two solid-fuel booster rockets (IUS and PAM). It arrived in the proximity of Jupiter on February 8, 1992, and, with gravitational assistance, was successfully redirected into a trajectory inclined 80 degrees to the plane of the ecliptic. Between 1994 and 1995, the spacecraft flew over the Sun's South Pole for the very first time, followed with a flyby of the North Pole. Between June 26 and November 5 of 1994, it circled the South Pole at a distance of 300 million kilometers, and a year later, between June 19 and September 29, 1995, flew over the North Pole. It's notable that Ulysses was between 2 to 2.3 AU from the Sun during these polar flights, or approximately twice the Earth–Sun distance. Both poles were circled again between 2000 and 2001, during a second orbit. The mission was subsequently extended until 2009, for a third polar flyby between 2007 and 2008 and to optimize scientific returns. As a result, Ulysses was able to measure the speed of the solar wind during a period of minimal solar activity (1994–1995) and maximum solar activity (2000–2001). Thanks to Ulysses, it was determined that interstellar atoms enter the heliosphere, become ionized during that passage and become part of the stream flow of the solar wind.

Ulysses depicted in its operational configuration, monitoring solar activity.

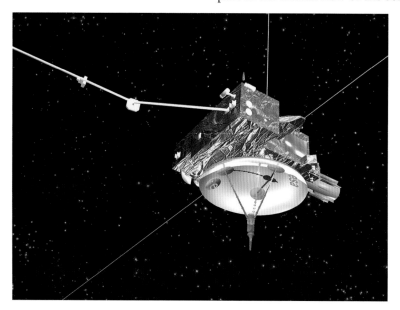

After three polar orbits and more than 17 years of operation covering a complete 11-year solar cycle, the Ulysses mission came to an end on July 1, 2008, due to a reduction in power of the thermo-electric generator, which in turn resulted in damage to the attitude-control system. Even though the mission is officially over, Ulysses continues to transmit useful data on the characteristics of the solar wind. ⊙

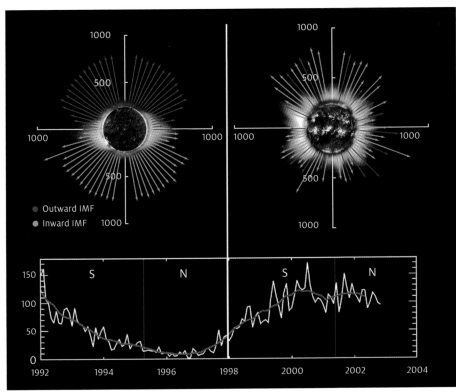

Solar-wind velocity measurements in kilometers per second are shown during weak (at left) and strong (at right) periods of solar activity, as obtained by the SWOOPS (Solar Wind Observations Over the Poles of the Sun) instrument on the Ulysses mission. The lower panel shows the correlation between the intensity of solar activity and the number of sunspots.

BELOW
Chronology of the third Ulysses orbit around the solar poles.

IMF = Interplanetary Magnetic Field

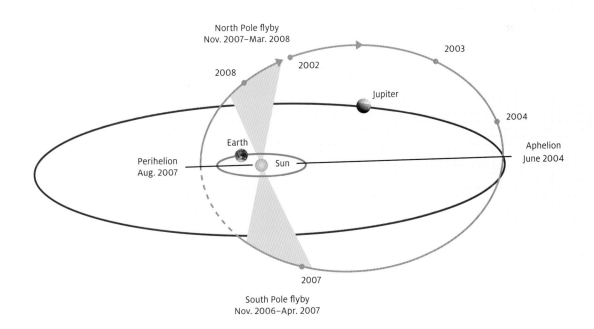

North Pole flyby
Nov. 2007–Mar. 2008

2008

2002

2003

Jupiter

2004

Earth

Perihelion
Aug. 2007

Sun

Aphelion
June 2004

2007

South Pole flyby
Nov. 2006–Apr. 2007

Wind

T HE SPACE PROBE WIND, built by Lockheed Martin for NASA, was launched on a Delta II 7925 rocket on November 1, 1994, to study the composition of solar wind and the magnetosphere. The Wind data that were obtained from the plane of the ecliptic were then compared to those of the Ulysses mission (see page 278), which were acquired at different solar latitudes. Wind was to occupy a halo orbit around the L_1 point; however due to the arrival of the SOHO and ACE probes (see pages 283 and 286), its positioning was delayed until 2004. Wind is currently in orbit and continues to transmit solar meteorological data from this region. The GGS mission (Global Geospace Science Mission), NASA's contribution to the ISTP (International Solar Terrestrial Program), used the Polar satellite, the Wind probe and the Japanese scientific satellite Geotail to study the behavior of the physical interaction between the Earth and the Sun. The second American component of the constellation, Polar, was launched on February 24, 1996, and placed in an elliptical orbit 51,200 kilometers above the Earth's North Pole and 138 kilometers above the South Pole. Polar and Wind studied the relation between the plasma emitted by the Sun and its interaction with Earth's magnetic poles, the magnetosphere and the ionosphere. ☉

MAIN COMPONENTS OF WIND

1. Plasma and radio-wave detector
2. Plasma and radio-wave antenna
3. Gamma-ray and X-ray detector
4. High-energy-particles detector
5. Doppler imager
6. Solar panel
7. Low-energy-particles detector
8. Gamma-ray spectrometer

The Wind probe is an integral part of the solar-storm detection network.

SOHO

THE SOHO (SOlar and Heliospheric Observatory) mission was launched from Cape Canaveral on December 2, 1995, by an Atlas IIAS rocket. Initially designed to operate for two years, SOHO still functions to this day. This is an international collaboration between the European Space Agency (ESA) and NASA, dedicated to studying the Sun. It has become the primary real-time information source of solar activity.

The spacecraft's total weight is 1,875 kilograms, including 1,640 kilograms for the probe and 235 kilograms of fuel. In order to place SOHO in the same relative position as Earth, it was launched into a halo orbit at the Lagrangian point L_1, a position in space between the Sun and the Earth where their respective gravitational forces balance out. This is a good strategic point for constant communication with Earth. Both Wind (see page 282) and ACE (see page 286) probes are also located at the L_1 point, at a distance of 1.5 million kilometers from Earth.

Final inspection of the SOHO assembly. This solar mission is a collaboration between ESA and NASA.

MAIN COMPONENTS OF SOHO

1. Spectrometric coronagraph
3. UV spectrometer
4. Energized-particle sensor
5. Solar-oscillation sensor
6. Doppler imaging
7. Solar panel
8. Low-gain antenna
9. Anisotropic solar-wind sensor
10. Solar-wind analyzer
11. UV coronal imaging
12. UV telescope
13. UV spectrometer

LEFT

The SOHO probe, currently located 1.5 million kilometers from Earth in orbit around the Lagrangian L_1 point, transmits data on the internal structure of the Sun.

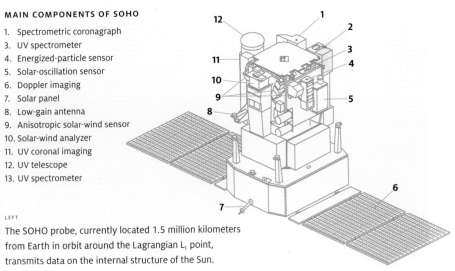

SOHO's UV telescope photographed the eruption of a giant prominence in the Sun's corona on September 14, 1999. The hottest regions of the Sun appear clearer in this image, which was taken through an ultraviolet filter.

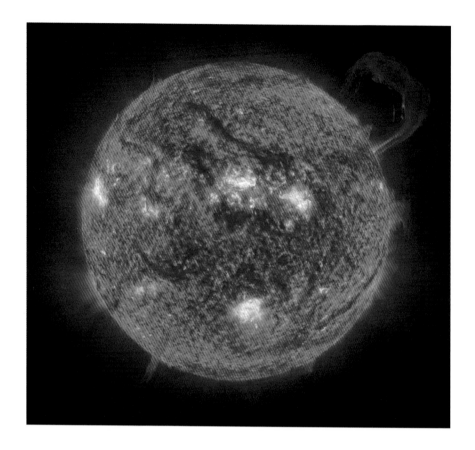

On June 24, 1998, an incident nearly triggered the end of SOHO. The probe began to spin on axis after a series of routine commands, and radio communications were interrupted. In the absence of thermal controls, temperatures fluctuate quickly in space, oscillating between −150°C in the shade and +200°C in the Sun. This can cause major thermal and mechanical stress on the probe's structure and instrumentation. Fortunately, SOHO was located by radar on July 23, when radar signals sent by the Arecibo radio telescope in Puerto Rico to find the probe were sent back and detected by the 70-meter antenna of NASA's Deep Space Network. Once telemetry resumed, the frozen hydrazine reservoirs were gradually defrosted. The spacecraft's engines were fired on September 16, to regain control of the probe's orientation in space and finally stabilize it. The mission resumed its normal course on October 24 without any damage to the instrumentation.

The 12 SOHO instruments measure the physical properties of the solar atmosphere, especially the chromosphere, the transitional zone and the corona (with two coronagraphs). Other devices monitor the solar

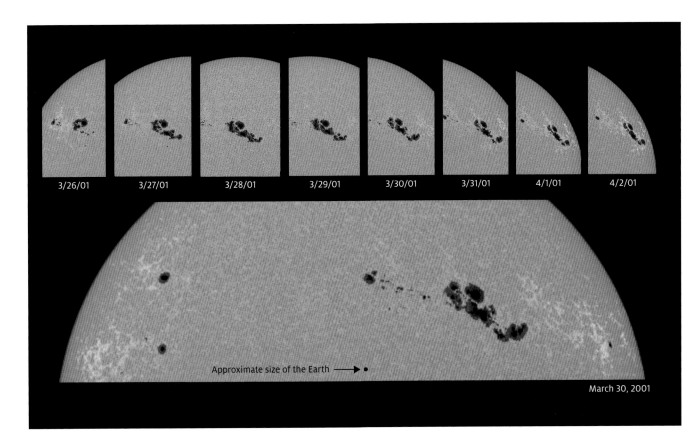

3/26/01   3/27/01   3/28/01   3/29/01   3/30/01   3/31/01   4/1/01   4/2/01

Approximate size of the Earth ⟶ •

March 30, 2001

wind in the L_1 region, while the Sun's internal structure is studied through helioseismology, using vibration sensors and by measuring oscillations in the Sun's interior. SOHO has become quite an astronomical celebrity, mainly due to the millions of images it has provided and the billions of measurements that have revolutionized our knowledge of the Sun. In addition, by constantly monitoring the ejection of coronal masses, SOHO has had a secondary benefit: it has discovered a large number of comets. In fact, more than half of all known comets have been discovered by the probe: in 2010 it discovered its 2,000th comet! This was made on December 26, 2010, by Michael Kusiak, an astronomy student in Krakow, Poland. This young scientist has been credited with the discovery of more than 100 comets. ☉

The largest array of sunspots ever observed. It was the source of several coronal mass ejections, including one that generated the largest blast of X-rays registered in 25 years.

BELOW

Comparison of solar activity over a three-year period, from the beginning of 1997 to the end of 1999, as documented by SOHO. This fell within an 11-year cycle that had seen the Sun attaining maximum activity in the year 2000.

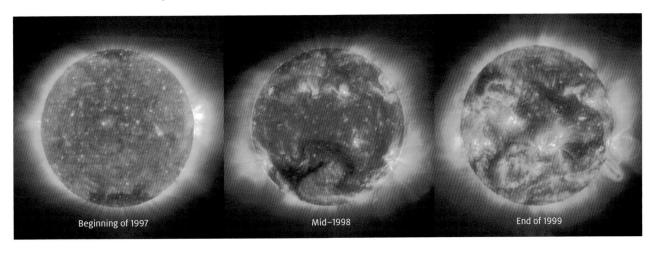

Beginning of 1997   Mid–1998   End of 1999

ACE

Nasa's still-active ace (Advanced Composition Mission) probe was launched from Cape Canaveral on August 25, 1997, by a Delta II 7920 rocket. The utility payload of the probe consists of nine instruments, four of which are high-resolution spectrometers to measure (with great precision) the elemental, isotopic and ionic composition of interplanetary space. ACE is also equipped with instruments that constantly monitor solar-wind flow and galactic emissions. The probe transmits its measurements to Earth in real time from a halo orbit facing the Sun at the Lagrangian point L_1, 1.5 million kilometers away from Earth. This location was chosen so that ACE is well outside of Earth's magnetosphere, to ensure that it samples original material coming from the Sun and the Milky Way. Along with other solar observatories, such as SOHO and Wind, Ace is part of a warning system used to forecast geomagnetic storms, with a one-hour advance alert time. The probe is equipped with enough fuel to maintain its position until the year 2024. ⊙

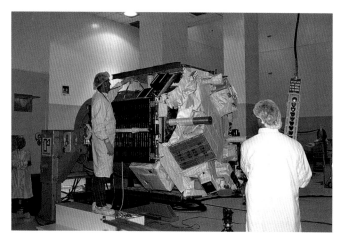

NASA technicians inspect the four solar panels of the ACE probe, which is part of the Explorer space program.

The data transmitted by ACE in real time are utilized by the Space Weather Prediction Center to forecast and provide early warnings of solar storms.

Memorable Magnetic Storms

When a coronal mass is ejected or a solar eruption is directed toward the Earth, an excess of electrically charged solar particles disrupts our magnetosphere. The solar-wind shock wave usually takes between 24 and 36 hours to reach the Earth, at a speed of several million kilometers per hour.

Between August 28 and September 2, 1859, several sunspots were observed, signs of increased solar activity. On September 1, a solar eruption was detected, followed by a gigantic coronal mass ejection, moving directly toward Earth. Moving at tremendous speed, the shockwave reached the Earth 18 hours later. Many telegraph lines in Europe and North America were bombarded by a torrent of charged particles and the resulting short circuit gave rise to several fires. Auroras were reported in Hawaii, Mexico and Italy, at latitudes much lower than normally observed.

On March 13, 1989, a severe magnetic storm, also caused by a coronal mass ejection, triggered a disastrous chain reaction of just a few seconds in Quebec's electric grid, thereby depriving six million residents of electricity for several hours. This storm also set off auroras sightings all the way to Texas.

MAIN COMPONENTS OF ACE

1. Isotopic spectrometer
2. Solar-ion analyzer
3. Electron and proton spectrometer
4. Mass spectrometer (H, He)
5. Solar panel
6. Solar-wind spectrometer
7. Mass spectrometer (m> He)
8. Attitude-control module
9. Magnetometer
10. Charged-particle spectrometer
11. Spectrometric telescopes

Genesis

AFTER THE APOLLO PROGRAM, Genesis was the first NASA mission seeking to return physical samples from space. The name of this mission, Genesis, refers to its scientific objective: to provide a better understanding of the conditions during the early formation of the Sun and solar system. Designed and built by Lockheed Martin for Jet Propulsion Laboratory (JPL), the Genesis probe was sent to the Lagrangian point L_1. Its mission was to collect samples of the solar wind for several years and then return them to Earth via an atmospheric re-entry capsule, for more detailed analysis. Since the probe was to be placed at L_1, at a distance of 1.5 million kilometers, and outside Earth's magnetosphere, the solar-wind samples could be considered "pure." On August 8, 2001, Genesis was launched by a Delta II 7326 rocket from Cape Canaveral. After an initial complex trajectory, Genesis orbited at L_1 from December 2001 to April 2004.

BELOW, LEFT
Assembly facility for the Genesis probe at NASA's Jet Propulsion Laboratory site in California.

BELOW, RIGHT
Each collector was composed of a mosaic of fine wafers, made of silicon, gold, sapphire and diamond.

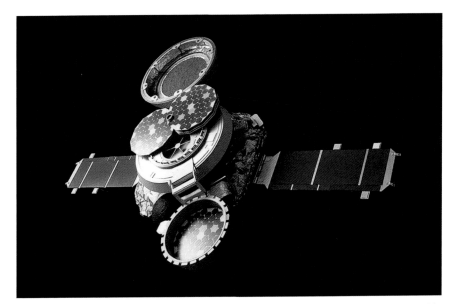

Artist's rendition of Genesis with its collectors deployed to capture solar-wind particles beyond the Earth's magnetosphere, at the Lagrangian point L$_1$.

During this time, the probe's three collecting discs were sequentially exposed to collect samples of solar wind at different stages of solar activity. The low-density materials in the collectors were selected to slow the flow of solar-wind components so as to collect and preserve them with minimal impact. The collectors, each the size of a bicycle wheel, consisted of arrays of semi-conductor wafers composed of silicon, gold, sapphire, diamond-like film and other materials. Genesis also carried three other instruments: an ion counter to register the speed, density, temperature and composition of the solar winds; an electron counter; and an ion concentrator to assemble and separate the various solar-wind elements, such as oxygen and nitrogen, into small targets.

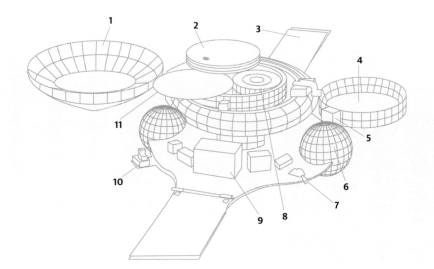

MAIN COMPONENTS OF GENESIS
(in collecting configuration)

1. Return capsule
2. Collector battery
3. Solar panel
4. Collector's cover
5. Star trackers
6. Fuel reservoir
7. Attitude-control thruster
8. Concentrator
9. Control module and processor
10. Ion detector
11. Deployed collector

One of the many in-flight recovery rehearsals for the Genesis capsule and its precious collection of extraterrestrial particles.

Due to technical problems during its descent, the capsule crashed violently into the Utah desert.

After being exposed, the collectors were enclosed in a hermetically sealed compartment to safeguard their return to Earth in the re-entry capsule. Since the collectors were too fragile for a traditional landing, the re-entry capsule was to have been recovered in flight, first being slowed by a parachute and then by a parafoil. Two helicopters would try to capture the parafoil in flight with a long grapnel. If, by chance, the helicopters missed their target, the parafoil would then further slow the capsule to a reasonable landing speed of 15 kilometers per hour. Although many trial runs predicted a successful outcome of the operation, it was not to be.

On September 8, 2004, Genesis began its return with a successful atmospheric re-entry, ready for a helicopter capture over the Utah desert. Unfortunately, the parachute failed to deploy 33 kilometers above ground, as planned, and the capsule crashed violently without braking at 311 kilometers per hour. The capsule suffered major structural damage and the precious samples were contaminated by surface dirt. The salvage team was quick to realize that the lubricants and other components of the capsule itself would complicate the decontamination efforts. Fortunately, the team was able to salvage the scientific goal of the mission through the enormous task of selecting and analyzing thousands of collector fragments.

A NASA inquiry commission investigated the causes of the crash. On October 14, 2004, it concluded that a miniature accelerometer had been installed improperly. Since this device controlled the sequence of parachute deployment, Genesis' fate had been sealed from the very onset of the mission. ⊙

TRAJECTORY OF THE GENESIS MISSION INTO A HALO ORBIT, FOLLOWED BY ITS RETURN TO EARTH

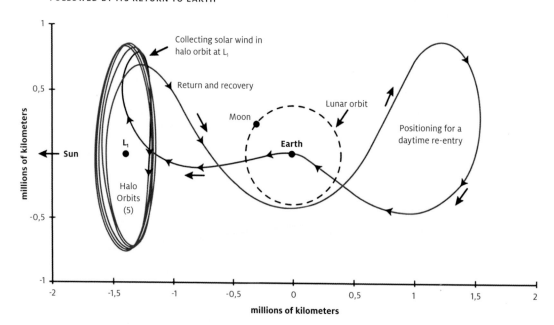

STEREO

NASA'S STEREO MISSION (Solar TErrestrial RElations Observatory) consists of two identical space probes in heliocentric orbits, one slightly ahead and the other slightly behind the Earth in relation to the Sun. By placing STEREO-A (Ahead) and STEREO-B (Behind) about 60 degrees apart, they are in optimal locations to obtain, for the first time, accurate stereoscopic views of solar events, especially coronal mass ejections. The two spacecraft were launched in tandem by a Delta II-7925 rocket, and assumed indirect trajectories with several gravitational boosts from the Moon, which made it possible to place each probe into a different orbit. STEREO-A was placed closer to the Sun, in an orbit that takes 346 days to complete (19 days faster than the Earth). STEREO-B was placed farther out from the Sun, in an orbit of 388 days, 23 days longer.

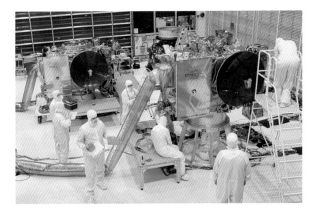

Parallel assembly of the two STEREO probes at NASA's Goddard Space Flight Center.

Double encapsulation of the two probes in the Delta II-7925 launcher.

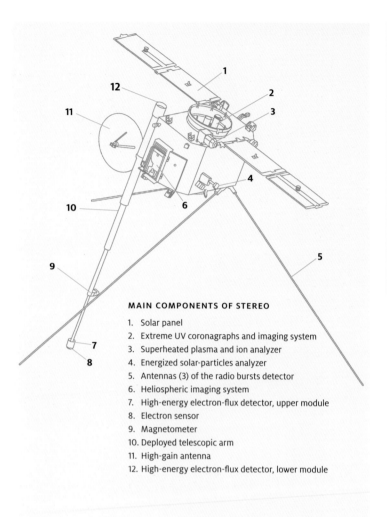

MAIN COMPONENTS OF STEREO

1. Solar panel
2. Extreme UV coronagraphs and imaging system
3. Superheated plasma and ion analyzer
4. Energized solar-particles analyzer
5. Antennas (3) of the radio bursts detector
6. Heliospheric imaging system
7. High-energy electron-flux detector, upper module
8. Electron sensor
9. Magnetometer
10. Deployed telescopic arm
11. High-gain antenna
12. High-energy electron-flux detector, lower module

Artist's rendition of the STEREO probes in a terrestrial orbit formation, before being separated to two different sides of the Earth.

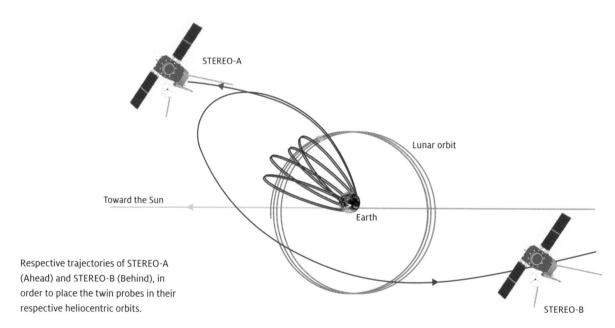

Respective trajectories of STEREO-A (Ahead) and STEREO-B (Behind), in order to place the twin probes in their respective heliocentric orbits.

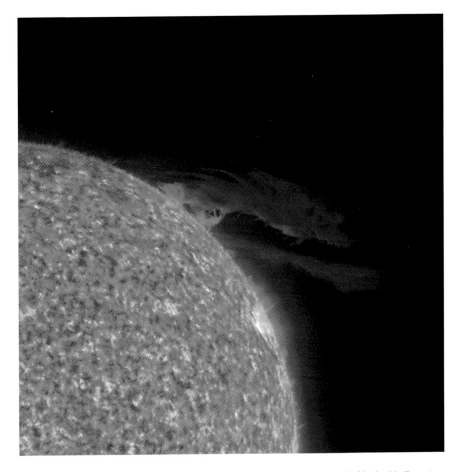

On September 29, 2008, STEREO-A observed this solar prominence, composed of ionized helium at 60,000°C. It flapped like a flag for several hours before breaking up and dispersing in space as part of the solar wind.

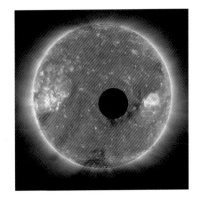

A STEREO image of our Moon transiting in front of the Sun.

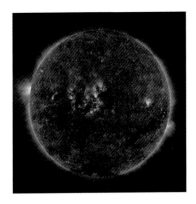

One of the first snapshots of the solar disk transmitted by the STEREO mission, on December 6, 2006. It was imaged by the SECCHI telescope in ultraviolet light.

STEREO-A and STEREO-B were separated at an angle of 180 degrees in February of 2011, an ideal configuration for observing the two sides of the Sun simultaneously. This has provided a complete 3D view of our home star for the first time. Each probe is equipped with five telescopes with digital cameras: one for imaging the Sun in the extreme ultraviolet light; two types of coronagraphs to observe the interior and exterior corona, and two to monitor the heliosphere between the Sun and Earth. In addition, multiple sensors measure solar radio emissions, energized-particle flows, the magnetic field and the solar wind. The two probes and all on-board instruments are functioning perfectly and continue to transmit data and amazing images of the Sun to this day. ☉

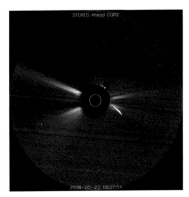

STEREO-A captured the last moments of a kamikaze comet as it disintegrated and disappeared into the solar furnace in May of 2008.

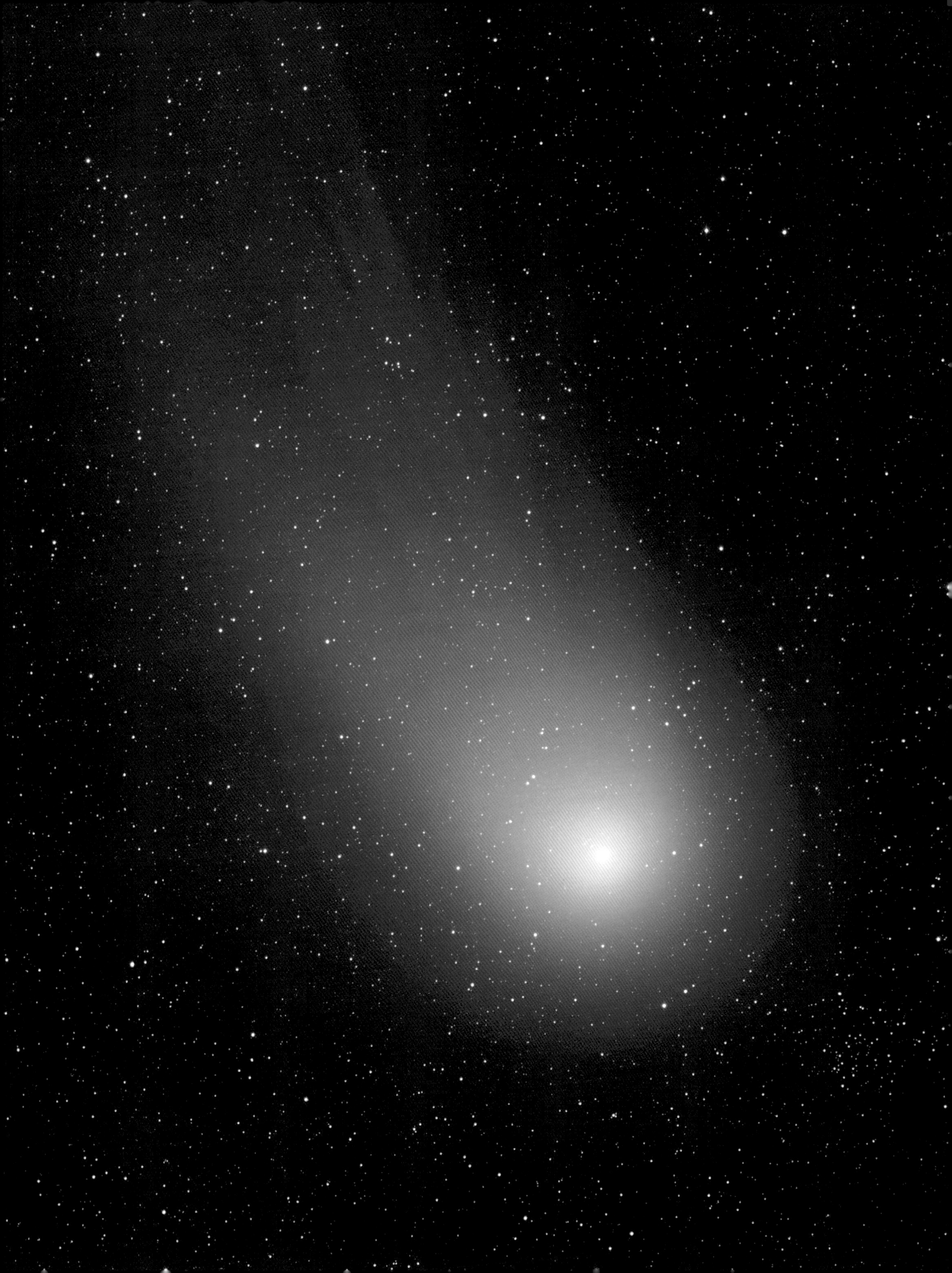

Comets

F ROM TIME IMMEMORIAL, the passage of a comet has been an unforgettable sight for those lucky enough to see it. Such events were long considered signs of either good or bad omens, depending on the historic events at the time. Comets are among the smallest solar-system objects that orbit the Sun. They have nuclei ranging in size from a few meters to a few tens of kilometers. From the nucleus emanates a coma and an immense bright tail of gas and dust, which can reach up to 150 million kilometers in length. The nucleus consists of a heap of rocks, ice, solid-state gas and dust that formed at the very beginning of the solar system. When a comet nears the Sun, radiation causes vaporization of water, gases and other volatile materials carrying along grains of dust in the comet. Both the freed gases and the dust form an atmosphere (coma) around the nucleus and a trail (tail) in the direction opposite to the solar radiation's pressure. The coma and the tail become visible from Earth when a comet enters the inner solar system, with the comet's dust reflecting sunlight and the gases becoming visible as they are ionized by the Sun's ultraviolet radiation.

The dust shed by a comet forms a large cloud in its wake. When the Earth crosses this cometary debris, dust grains burn up in the upper atmosphere and leave ionized trails in the sky, which we call a meteor shower. The annual, mid-August Perseid meteor shower coincides with the Earth's crossing the cloud left by Comet Swift-Tuttle.

Comets originate mainly from two sources: the Kuiper Belt, which extends from Neptune's orbit out to 55 AU from the Sun, and the Oort Cloud, farther out, which surrounds the solar system at more than 50,000 AU from the Sun.

The comet of 1858, discovered by Italian astronomer Giovanni Battista Donati, was the most spectacular comet of the 19th century and the first to be photographed.

OPPOSITE PAGE

The comet C/2001, also known as Comet NEAT. This image was taken by telescope from the Kitt Peak National Observatory near Tucson, Arizona.

This French caricature, dating back to 1857, illustrates the fear that often results from the passage of a comet in the sky.

BELOW

In 2007, Comet McNaught, the most brilliant in several decades, was visible to the naked eye in broad daylight in the southern hemisphere.

Comets arising from the Kuiper Belt have short-period elliptical orbits, lasting less than two years, while those originating from the Oort Cloud have long-period orbits, lasting thousands of years. In 1705, the English astronomer Edmund Halley was the first to conclude that comets witnessed in 1531, 1607 and 1682 were one and the same object. With a periodicity of 76 years, the comet (since named Halley's Comet) returned in 1758, as he had predicted. The next return of Halley's Comet is predicted for 2061. More than 4,000 comets have been identified, including around 500 with a predictable short-period orbit. The majority are comets of the Kreutz group, which pass close to the Sun and are known as sungrazers (see comet discovery by SOHO on page 285). In 1888 the German astronomer Heinrich Kreutz hypothesized that sungrazing comets are related, originating during the fragmentation of a single comet's nucleus in 1106. Several spectacular giant comets — such as those of 1843, 1882 and the Ikeya-Seki of 1965 — belong to this group.

The Great Comet of 1843 was so bright that it was visible in daylight from the southern hemisphere. The tail of this remarkable sungrazer spread over more than two AU at its perihelion.

During the early stages of formation of the solar system, the frequency of collisions between comets and planets was much greater than it is today. This makes it quite likely that the large quantities of water found on Earth came originally from comets. It is quite rare to witness the collision between a comet and a planet. However, that is exactly what happened in July of 1994, when Comet Shoemaker-Levy 9 disintegrated under the combined gravitational forces of the Sun and Jupiter. The resultant fragments, each about the size of a mountain, crashed into Jupiter with an explosive force equal to that of 40 million megatons of TNT, more than the force of all existing nuclear bombs combined.

With the onset of the space age, our knowledge of comets took both a quantitative and qualitative leap. The in situ observations of comet nuclei and the analysis of their tails by several robotic probes have greatly helped to clarify their origins and composition. The European Rosetta mission will attempt to land on the nucleus of comet 67P/Churyumov-Gerasimenko in 2014.

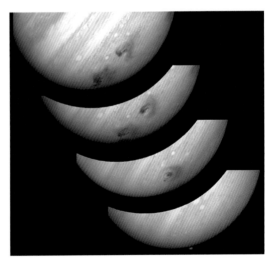

A rare astronomical event was observed in July 1994: Comet Shoemaker-Levy 9 was torn into fragments by Jupiter's gravity and crashed into the giant planet in a series of formidable collisions, which left scars in Jupiter's atmosphere for many months.

1984

Vega 1 and 2

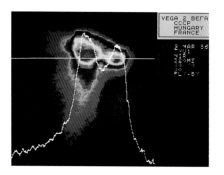

On March 9, 1986, the Soviet probe Vega 2 approached Halley's Comet at a distance of 8,030 kilometers and was able to transmit 700 images during its brief encounter.

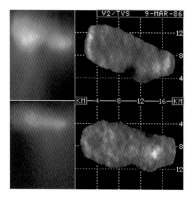

Image of Halley's Comet topography, extrapolated from Vega 2 images by a procedure known as background noise reduction.

OLLOWING THEIR ENCOUNTER with Venus in June 1985 (see page 116), the Vega 1 and Vega 2 probes were placed on a trajectory to intercept Halley's Comet. During the comet's 1986 passage, an international flotilla of five probes — the Soviet Vega, the European Giotto, the Japanese Suisei and Sakigake — was nicknamed "Halley's Armada." The American participation in the study of Halley's Comet was limited to long-distance observations by ICE (see page 276). The first images of the comet's nucleus were transmitted by Vega 1, beginning on March 4, 1986, which helped in adjusting Giotto's final trajectory (see page 301). Protected by a shield from the dust emitted by the comet's tail, the probes made their closest approach to the nucleus a few days later: Vega 1 on March 6, 1986, at 8, 890 kilometers and Vega 2 on March 9 at 8,030 kilometers. These close observing sessions lasted three hours. The images showed two 14 kilometer-long bright gas jets emanating from the comet's dark nucleus, which was rotating with a period of 53 hours. In total, 1,500 images of Halley's Comet were sent by Vega 1 and Vega 2. The impact of cometary particles with the probes significantly reduced the output of their solar panels (down 40 percent for Vega 1 and 80 percent for Vega 2), but none of their instruments were damaged. Analysis of the comet's dust particles indicated that they are composed of the same elements as chondrites, primitive meteorites found on Earth. The Vega probes were also able to measure the temperature of the comet's nucleus at between 27°C and 130°C, rather surprising for a frozen object. Now that their missions are completed, the probes travel in silence on their heliocentric orbit.

Sakigake and Suisei

THE JAPANESE ISAS (Institute of Space and Astronautical Science, now part of JAXA, the Japan Aerospace Exploration Agency) designed its Sakigake ("pioneer") mission as a prototype, primarily to test the country's astronautical interplanetary technology, especially long-distance radio communications. Sakigake was the first interplanetary probe launched by a country other than the United States and the Soviet Union. It was launched on January 7, 1985, from the Kagoshima Space Center by a four-stage M3S-II rocket on a direct path to Halley's Comet. This small, 138-kilogram spacecraft was equipped with only three instruments. The main scientific goal was to measure the interaction between the comet and both the solar wind and interplanetary magnetic fields. On March 11, 1986, it successfully approached the comet to within 7 million kilometers, followed by a return to Earth on January 8, 1992, and a flyby at a distance 89,000 kilometers, making it the first Japanese spacecraft to accomplish a planetary flyby.

The Sakigake and Suisei probes had the same cylindrical shape.

MAIN COMPONENTS OF SAKIGAKE

1. High-gain antenna reflector
2. UV imaging
3. Attitude-control unit
4. Solar panel
5. Medium-gain antenna
6. Low-gain antenna
7. Charged-particle analyzer

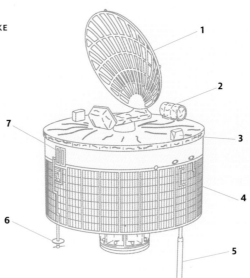

Halley's Comet, as it appeared during its 1910 passage.

Sakigake's mission was extended for potential encounters with comets Honda-Mrkos-Pajdusakova in 1996 and Giacobini-Zinner in 1998. Unfortunately, after loss of radio contact with Sakigake in November 1995, the mission was aborted.

On August 18, 1985, Susei ("comet" in Japanese; originally named Planet-A) was propelled toward Halley's Comet by an M3S-II rocket. Identical in structure to Sakigake, this probe was equipped with an ultraviolet imaging system to study the huge hydrogen corona around the comet. Thanks to a trajectory-correcting maneuver, Susei came within 151,000 kilometers of the comet on March 8, 1986, and was able to analyze its coma and its 20 million-kilometer hydrogen tail. Despite its distance from the target, the spacecraft was struck by two cometary grains, no more than a millimeter in diameter. ISAS decided to send Susei toward comet Giacobini-Zinner for a November 1998 rendezvous. This was a complex trajectory, requiring several course corrections and a gravitational assist from Earth. Unfortunately, the probe ran out of fuel to carry out this mission and had to be deactivated on February 28, 1998. ✌

Giotto

T HE GIOTTO PROBE was ESA's first interplanetary spacecraft. Its mission was a close flyby of Halley's Comet's nucleus during its 1986 passage. The comet's next passage won't occur until the year 2061. The spacecraft's name pays homage to the Florentine painter, Giotto di Bondone (see page 305), who painted Halley's Comet in a work called *The Adoration of the Magi*. Originally Giotto was to be accompanied by a partner probe from NASA, but the space agency was encountering budgetary setbacks at the time and the project was cancelled. NASA had planned to proceed with observations of Halley's Comet using telescopes aboard the space shuttle, but unfortunately, that plan was shelved because of the Challenger disaster.

Preparation of the Giotto probe at the Intespace site in Toulouse, France.

Giotto undergoing a rotational test in a solar radiation chamber, simulating conditions in interplanetary space.

On July 2, 1985, the first day of the official launch-window opportunity, the Giotto mission is successfully propelled by an Ariane-1 rocket at Kourou in French Guyana.

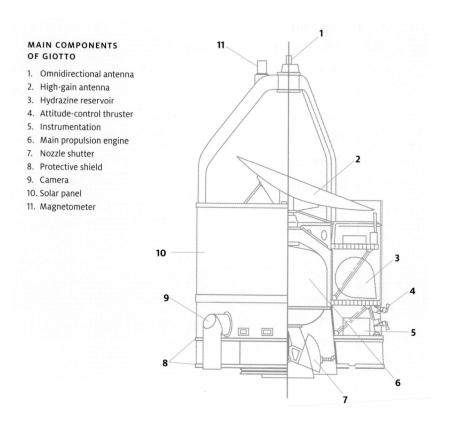

MAIN COMPONENTS OF GIOTTO

1. Omnidirectional antenna
2. High-gain antenna
3. Hydrazine reservoir
4. Attitude-control thruster
5. Instrumentation
6. Main propulsion engine
7. Nozzle shutter
8. Protective shield
9. Camera
10. Solar panel
11. Magnetometer

TRAJECTORIES OF EARTH, THE GIOTTO PROBE AND HALLEY'S COMET

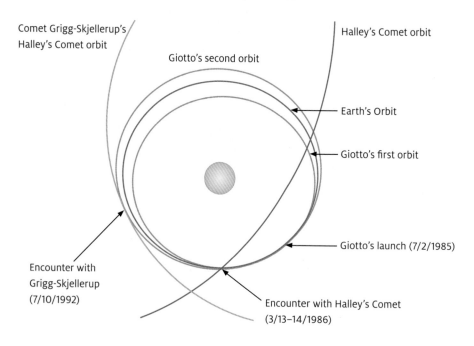

Comet Grigg-Skjellerup's
Halley's Comet orbit

Giotto's second orbit

Halley's Comet orbit

Earth's Orbit

Giotto's first orbit

Giotto's launch (7/2/1985)

Encounter with
Grigg-Skjellerup
(7/10/1992)

Encounter with Halley's Comet
(3/13–14/1986)

The Giotto spacecraft was similar in structure to the British Aerospace GEOS research satellite, but was equipped with a special aluminum and Kevlar shield to protect it from bombardments by cometary dust grains. In addition, it was equipped with a color digital camera to study and image the comet's nucleus, and mass spectrometers to characterize the chemical compositions of the comet's coma and tail. Several dust and plasma sensors and a photopolarimeter rounded off the array of scientific instruments on board the 583-kilogram probe.

The Giotto mission took off on July 2, 1985, launched by an Ariane I rocket from the European Space Center at Kourou in French Guyana. The probe was first placed into a geostationary orbit, then an ABM (Apogee Boost Motor) booster rocket propelled it toward its distant encounter with Halley's Comet. Guided by the ESA control center in Darmstadt, Germany, and thanks to the precise data furnished by the two Vega probes, Giotto was able to approach the nucleus of Halley's Comet to within a distance of only 600 kilometers on March 14, 1986. Incredibly, the probe was able to withstand the impact of some 12,000 recorded pieces of cometary debris! Darker than soot, the comet nucleus had an irregular, elongated shape, measuring 15 kilometers in length and 7 to 10 kilometers in width. At least three vapor jets were detected from the comet's surface. The analysis of comet materials indicates that it is 4.5 billion years old and similar in chemical composition to the primordial Sun. The comet was ejecting three tons of material per second into space, mostly water and carbon monoxide. One particularly violent impact caused Giotto to deviate from its trajectory and lose orientation, thereby leaving it exposed for at least 30 minutes. Another impact destroyed the camera — but this was after it had already transmitted several spectacular images of the nucleus. The probe finally reset its orientation and continued along its course.

On July 2, 1990, Giotto passed by the Earth at a distance of 22,000 kilometers for a gravitational boost that redirected it toward comet Grigg-Skjellerup. It reached its destination on July 10, 1992, passing within 200 kilometers. This mission, considered an encouraging success for the young ESA, was officially declared over on July 23, 1992.

Artist's rendition of Giotto flying by Halley's Comet at a safe distance.

Geographic distribution of the radio antenna network used during the Giotto mission. The numbers correspond to the diameter of antennas in meters. DSN = Deep Space Network.

BELOW

Timed-sequence images of Giotto's approach phase to Halley's Comet.

EN BAS OF PAGE

On March 14, 1986, Giotto approached the comet's nucleus to within an astonishingly close distance of 600 kilometers.

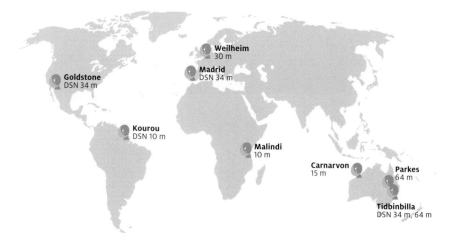

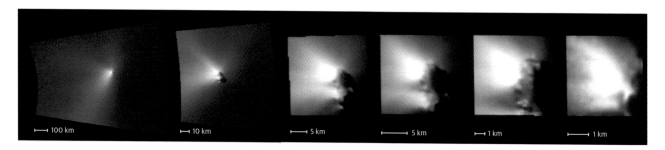

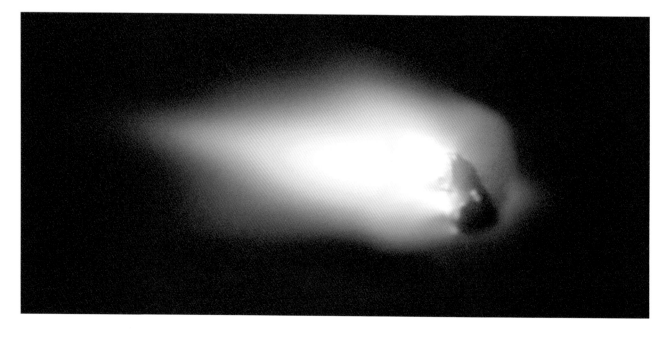

GIOTTO

(circa 1267-1337)

GIOTTO

Giotto di Bondone, or simply Giotto, was a painter, sculptor and architect from Florence, Italy. Today he is considered one of the major Italian pre-Renaissance artists. He is especially well known for his frescoes, which he painted on walls in several churches in Florence, Assisi and Padua. The series of Giotto's frescoes in the Scrovegni Chapel of Padua, a masterpiece he painted in 1305, is among the world's greatest art treasures. Since Halley's Comet appeared in 1301, it is entirely possible that it was the inspiration for Giotto to represent the star of Bethlehem as a comet in his painting, *The Adoration of the Magi*. His representational painting was thought to be quite innovative, because he broke with the more stylized Byzantine method, which had up to that time greatly influenced other artists in the peninsula.

One particular anecdote from Giotto's circle of friends relates that Pope Benedict XI asked Giotto, through an intermediary, of proof of his talent. In response, Giotto drew a perfect circle freehand in red ink, on a sheet of paper. Impressed by his genius, the pope then commissioned several works from him.

Deep Space 1

BELOW, RIGHT

At full capacity, Deep Space 1 ion thrusters can lift a sheet of paper. However, this miniscule force can operate in this manner for several years, surpassing the capacities of chemical engines for interplanetary distances.

BELOW, LEFT

The Deep Space 1 mission tested a number of advanced technologies in space, including a transponder and an ion propulsion engine.

T HE DEEP SPACE 1 MISSION inaugurated NASA's New Millennium Program. Under the direction of the Jet Propulsion Laboratory (JPL) from 1995 to 2008, this program was charged with testing new, light, economical and reliable technologies that could be applied to several exploratory probes in the future. The goal of the Deep Space 1 mission to asteroid Braille and Comet Borelly was to validate several advanced technologies while visiting these two objects. For the very first time, the NSTAR electrostatic ion thruster was used as the propulsion system, as well as a new autonomic navigation system, AutoNav.

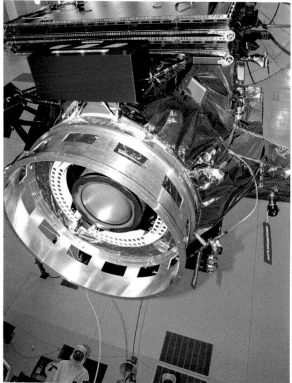

This system used images of celestial objects taken by a star tracker as navigational reference points, thereby requiring fewer trajectory adjustments. A new artificial intelligence program, Remote Agent, was tested to diagnose and correct any on-board malfunction, without human intervention. An innovative radio communication system, Beacon Monitor, was based on codified signals monitoring the overall status of the spacecraft. This was used on an experimental basis, to reduce the amount of work required to interpret telemetric signals on Earth during long interplanetary voyages.

The solar panels of Deep Space I use a sunray concentration method known as SCARLET (Solar Concentrator Arrays with Refractive Linear Element Technology) to produce more electrical energy (2,500 watts) over a smaller surface area. Several miniaturized pieces of equipment were also tested, including the digital imaging system MICAS (Miniature Integrated CAmera Spectrometer), the plasma sensor PEPE (Plasma Experiment for Planetary Exploration) and the SDST (Small Deep Space Transponder).

Deep Space I was launched on October 24, 1998, by a Delta II 7326 rocket and targeted at the small asteroid Braille (registered as 1992 KD). Mission control immediately encountered problems with the ion thrust engine, which shut down after only four and a half minutes. Thankfully, repeated efforts to reactivate succeeded and fixed the problem for good. Another hitch in the mission was loss of the star-tracker camera, which left the mission without visual navigation assistance. Finally, the spectrometer's ultraviolet channel of the MICAS module failed. The flyby of Braille on July 29, 1999, was only a partial success. Deep Space I was supposed to approach Braille as close as 240 meters, at a speed of 56,000 kilometers per hour. Unfortunately, a glitch in the navigational program and other technical difficulties caused it to pass the asteroid at a distance of 26 kilometers.

Deep Space I in cruising configuration.

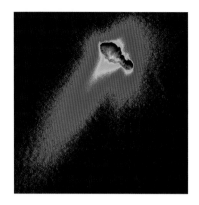

This composite image of Comet Borelly was obtained by Deep Space 1 from a distance of 4,800 kilometers. The false colors delineate the brightness of the coma and the dust ejected from the comet's nucleus.

In addition, the asteroid proved too dim for the AutoNav system to send the camera a focusing command, making the few images of Braille that were recorded out of focus. Fortunately, the flyby of Comet Borelli from a distance of 2,200 kilometers on September 21, 2001, was a total success, resulting in the best images of a comet's nucleus up to that time. Even though Deep Space I was not equipped with a protective shield, it survived encounters with cometary debris without damage, because the comet jets were not pointing toward it. NASA decided to suspend the mission on December 18, 2001, since the probe was running low on attitude-control fuel. The Deep Space I mission had a positive impact, due to the number of technological advances it provided that could be applied to future missions. For example, more recent NASA probes now use an optical navigation system similar to AutoNav for autonomous navigation. Another example of new technology is the ion thrust engine derived from NSTAR, now in use on the Dawn mission (see page 340).

MAIN COMPONENTS OF DEEP SPACE 1

1. High-gain antenna
2. Plasma analyzer
3. Solar screen for spectrometric imaging
4. Voltage conversion unit
5. Spectrometric imaging system
6. Electric distribution unit
7. Electronic module
8. Remote and diagnostic sensors
9. Propulsion module
10. Inertia reference unit
11. Star tracker
12. Solar panels (folded)

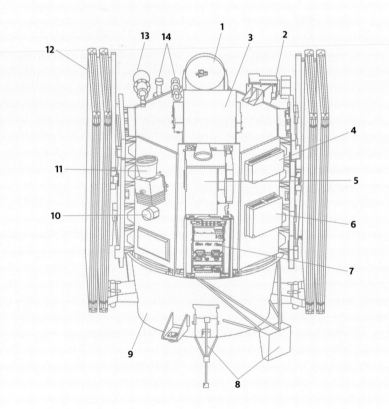

Stardust

THE MAIN OBJECTIVE OF NASA's Stardust mission was to study the composition of Comet Wild 2, named after the Swiss astronomer Paul Wild (pronounced "Vilt") who discovered it in 1978. In addition, it was the first space mission to bring back material from an extraterrestrial object since Luna 24 in 1976 (see page 58). After traveling a loop of three billion kilometers, Stardust returned a capsule to Earth containing samples of the coma from Comet Wild 2, as well as particles of interstellar dust.

An unusual feature of the Stardust mission was the nature of its sample collector. In order to prevent the destructive high-speed impact of sample particles on a solid surface, they were collected and retained in aerogel cells, a spongy silicon-based material of unusually low density.

A Lockheed Martin technician conducts final inspections of the Stardust probe prior to its encapsulation.

LEFT

The Stardust interstellar dust and comet particle collector is composed of aerogel cells to soften any impact.

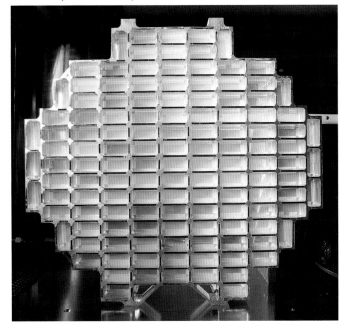

Dr. Peter Tsou, of the Jet Propulsion Laboratory, demonstrates the remarkable mechanical properties of aerogel, a very low-density insulating material.

BELOW
Artistic rendition of the encounter between Stardust and Comet Wild 2.

In addition, a miniature mass spectrometer, similar to that aboard the Giotto probe (see page 301), analyzed the molecular composition of dust particles captured in real time. Several Whipple screens protected the solar panels and the probe's body during its encounter with the comet.

On February 7, 1999, Stardust was launched on a Delta II rocket. This type of launch vehicle did not provide enough lift to propel the spacecraft on a direct trajectory to Comet Wild 2. Consequently, the probe first entered a heliocentric orbit and then received a gravitational push from the Earth on January 15, 2001. Some 17 hours later, it flew by the Moon at a distance of 108,000 kilometers and took 23 images in order to calibrate its camera. On its second orbit, the spacecraft flew by and photographed the asteroid Anne Frank, from a distance of 3,300 kilometers.

MAIN COMPONENTS OF STARDUST

1. High-gain antenna
2. Medium-gain antenna
3. Low-gain antenna
4. Interstellar and comet-dust collector (deployed)
5. Recovery capsule (open)
6. Interstellar and comet-dust analyzer
7. Launch adapter
8. Protective shields
9. Solar panel

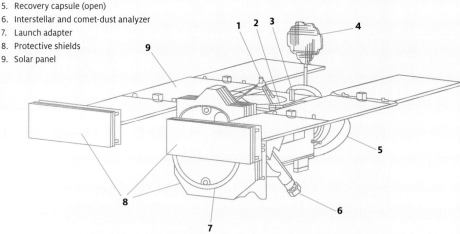

During its third orbit, Stardust crossed the comet Wild 2 on January 2, 2004. The probe transmitted detailed images of the comet's nucleus and was able to collect samples of its coma.

During this voyage, the collector was deployed and pivoted in order to collect interstellar dust particles on its unused surface. Nearing Earth, Stardust jettisoned the recovery capsule. The latter penetrated into the atmosphere at a lightning speed of 12.9 kilometers per second (46,440 kilometers per hour!). On January 15, 2006, spectators witnessed the descent of a ball of fire, followed by a sonic boom above the Great Salt Lake in Utah. The capsule and its parachute were blown away by a few kilometers; however, the soft landing occurred at the interior of the target zone with only a one-minute delay. The capsule was then transferred into a clean white room at the Johnson Space Center in Houston, Texas. They used the same procedure that was followed with the lunar soil samples during the Apollo era. The physical and chemical analyses of the samples is ongoing: there are millions of grains of comet dust, varying in size from a few microns to 1 millimeter in diameter, and 45 interstellar particles. The patient examination of microscopic impacts on aerogel images was done with the help of thousands of volunteers, who logged on to Stardust@Home on their personal computers.

Composite image of Comet Wild 2, as seen by Stardust on January 2, 2004. The diameter of the nucleus is 5 kilometers.

TRAJECTORIES OF THE STARDUST PROBE AND COMET WILD-2

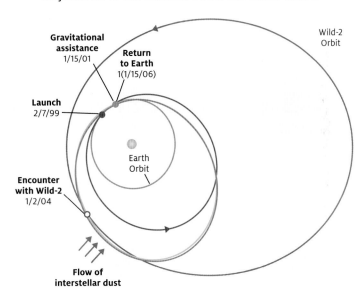

Gravitational assistance
1/15/01

Return to Earth
1(1/15/06)

Launch
2/7/99

Encounter with Wild-2
1/2/04

Flow of interstellar dust

Wild-2 Orbit

Earth Orbit

Data published to date have shown that nitrogenous organic compounds and hydrocarbons are present in the comet's material, which further support the hypothesis that the Earth may have been seeded by prebiotic molecules from comets.

Stardust was subsequently placed into sleeper mode in a heliocentric orbit. NASA approved an extension of the mission in July of 2007 and renamed it NExT (New Exploration of Tempel 1), with new plans for a flight to Comet Tempel 1 in 2011. It was to follow up the Deep Impact mission (see page 318), which was unable to observe the comet under good conditions after the comet was struck by the mission's impactor. The second comet flyby took place as planned on Valentine's Day, February 14, 2011. Stardust-NexT came to within 180 kilometers of the comet's nucleus and was able to image the crater left by the impactor of the Deep Impact mission.

The Stardust re-entry capsule landed at a military base in Utah on January 15, 2006 (left), and was then transferred to a special laboratory at the Johnson Space Center in Florida, for sample analysis (right).

A clearly visible micro-crater in one of the collector cells. Bottom: The track of a cometary particle impact, trapped in aerogel.

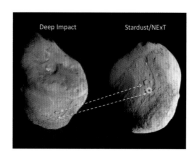

Comet Tempel 1, as seen by the probes Deep Impact and Stardust. Two craters, about 300 meters in diameter, help scientists locate the area hit by the Deep Impact impactor in July 2005 (red spot). Stardust approached the comet from a different angle on February 14, 2011.

Dr. Donald Brownlee, principal investigator of the Stardust mission, gives the victory sign after realizing that samples collection had been successful.

Rosetta

THE MAIN OBJECTIVE of the ESA's ambitious Rosetta mission, which is presently underway, is to visit Comet 67P/Churyumov-Gerasimenko, orbit it and deploy its lander, Philae, onto its surface. En route, Rosetta will fly by two asteroids. "Rosetta" refers to the Egyptian village where the French scientist, Jean-François Champollion (1790–1832), discovered the famed stone that was the key to interpreting ancient hieroglyphics. By analogy, this space mission hopes to unveil key elements that will shed some light on the origins of the solar system. "Philae" is the name of the island on the Nile where the obelisk was discovered that made it possible for Champollion to finally decode the ancient Egyptian writings.

The Rosetta probe was originally planned for launch in 2003, for an encounter with Comet 46P/Wirtanen in 2011, but launch problems and other delays led to a complete revision of the mission and its trajectory. Rosetta was launched on March 2, 2004, by the heavy lifter Arianne 5, from ESA's launch site at Kourou in French Guiana.

BELOW, LEFT
Inspection of Rosetta's giant solar panels at the Kourou Space Center facilities in French Guiana.

BELOW, RIGHT
The powerful European rocket, Ariane 5, launched Rosetta on a long mission that will last a decade (2004–2014).

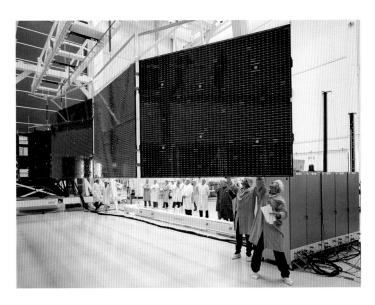

MAIN COMPONENTS OF ROSETTA

1. Low-gain antenna
2. Star trackers
3. Mass spectrometer
4. Comet-dust analyzer
5. Radio probe
6. Solar panel
7. Philae lander
8. Attitude-control thruster
9. Radio probe
10. High-gain antenna
11. OSIRIS imaging system
12. Langmuir probe (2)
13. Ion analyzer
14. Microwave telescope

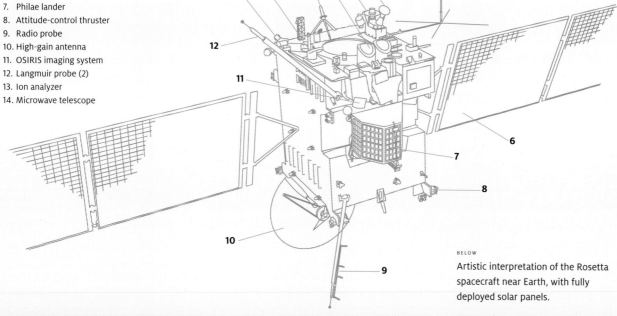

BELOW

Artistic interpretation of the Rosetta spacecraft near Earth, with fully deployed solar panels.

Despite the power of the Ariane 5 rocket, the mission will require several gravitational assists in order to reach comet 67P/Churyumov-Gerasimenko by 2014. The probe has already made two flybys of Earth (on March 5, 2005, and on November 13, 2007) and one past Mars on February 25, 2007. This very low flyby at only 250 kilometers above Mars was very risky, since the spacecraft could not use its solar panels while in the planet's shadow, nor communicate with Earth for fifteen long minutes. Mission control placed the probe in sleep mode during the flyby, and could not guarantee being able to reactivate it again. However, the acceleration maneuver was successful, and Rosetta continued on its path as planned. A third Earth flyby took place on November 13, 2009. Rosetta subsequently had a rendezvous with (and photographed) 2887 Steins, a small, 5-kilometer diameter, diamond-shaped asteroid. During this encounter, the probe's autonomous optical navigation and guidance system had a good trial run in preparation for its planned flyby of the comet. ✕

TRAJECTORY OF ROSETTA MISSION

1 Launch
3/2/04

2 Gravitational assistance
from Earth
3/4/05

3 Gravitational assistance
from Mars
2/25/07

4 Gravitational assistance
from Earth
11/13/07

5 Steins flyby
9/5/08

6 Gravitational assistance
from Earth
11/13/09

7 Lutetia flyby
7/10/10

8 Encounter with Comet
Churyumov-Gerasimenko
5/22/14

9 Landing on the comet
11/10/14

ORBITS

——— Earth

——— Mars

——— Steins Asteroid

——— Lutetia Asteroid

——— Comet Churyumov-Gerasimenko

ABOVE, LEFT

The planet Mars, as imaged from a distance of 240,000 kilometers by the Rosetta spacecraft OSIRIS camera.

ABOVE, RIGHT

The Philae lander's CIVA camera captured this image of Mars four minutes before the delicate gravitational boost, during Rosetta's flyby of the planet.

BELOW

Details of the harpoon that will anchor the Philae lander onto Comet 67P/Churyumov-Gerasimenko.

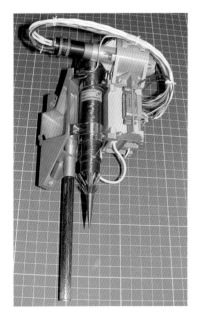

On July 10, 2010, Rosetta flew by Lutetia 21, a 140-kilometer diameter asteroid, and obtained some highly detailed images of it. The mission's orbiter is equipped with an array of sophisticated instruments to obtain data on the comet's nucleus and coma. OSIRIS (Optical, Spectroscopic and Infrared Remote Imaging System) has two cameras, one with a wide-angle and the other with a telephoto lens. Other equipment includes the spectral-imaging system VIRTIS (Visible and Infrared Thermal Imaging Spectrometer), which was used on Venus Express (see page 124), different types of optical spectrometers, a mass spectrometer and an atomic-powered microscope to examine cometary dust particles, captured on a silicon plaque. The 100-kilogram Philae lander is equipped with its own set of cameras, including a panoramic camera, CIVA (Comet Nucleus Infrared and Visible Analyzer), which was successfully tested during the dangerous flyby of Mars. Philae also carries with it several spectrometers for analysis of surface components of the comet's nucleus and APXS (Alpha Particle X-ray Spectrometer), which is similar to that on Mars Pathfinder (see page 166) and the Spirit and Opportunity rovers (see page 176). Rosetta's CONSERT (Comet Nucleus Sounding Experiment by Microwave Transmission) radar will permit deep probing of the interior of the comet's nucleus, and will maps its surface to select the best landing site. Plans are to place Rosetta in hibernation during its long voyage from July 2011 to January 2014, before its encounter maneuvers with the comet in May of 2014, and deployment of the Philae lander in November of that year. The mission is expected to end in December 2015, more than 4,000 days after its adventure began.

The Philae lander as it could appear on the comet's nucleus, if all goes according to the ESA plan.

This Rosetta image of Lutetia was taken from a distance of 3,162 kilometers. It supports the idea that this asteroid is a survivor of the violent events that gave rise to our solar system.

Rosetta came to within 800 kilometers of the small, diamond-shaped asteroid Steins.

2005 Deep Impact

BELOW, LEFT

The Deep Impact Mission was part of NASA's Discovery Program. The spacecraft was built by the American company Ball Aerospace & Technologies.

BELOW, RIGHT

On January 12, 2005, a Delta 2 rocket propelled the Deep Impact Mission toward Comet Tempel 1 at a velocity of 103,000 kilometers per hour, so that it could cross the 430-million-kilometer distance in 174 days.

THE CHALLENGE OF NASA'S Deep Impact Mission was to observe the effects of a collision with a comet in order to understand its composition. The short period (6.5 years) comet, 9P/Tempel 1, was the target selected for this daring experiment. By sheer coincidence, a science-fiction film with the same title, directed by Mimi Leder and released in 1998, tells the story of how the trajectory of an enormous comet was deflected so as to avoid a fatal collision with Earth.

**TRAJECTORIES OF DEEP IMPACT
AND COMET TEMPEL 1**

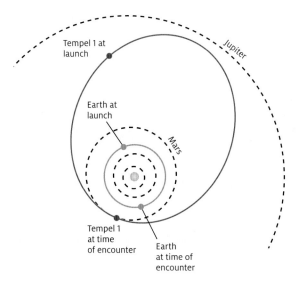

Tempel 1 at launch

Jupiter

Earth at launch

Mars

Tempel 1 at time of encounter

Earth at time of encounter

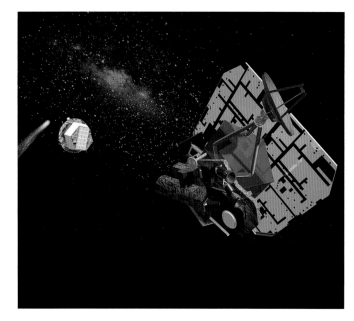

Artist's interpretation of the Deep Impact impactor on its way to Comet Tempel 1. The probe's solar panel plays an important protective role against the cometary dust particles.

MAIN COMPONENTS OF DEEP IMPACT

1. High-gain antenna
2. Low-gain antenna
3. Star tracker
4. Infrared spectrometer radiator
5. High-resolution imaging system and infrared spectrometer
6. Impactor
7. Medium-resolution imaging system
8. Solar-shield panel

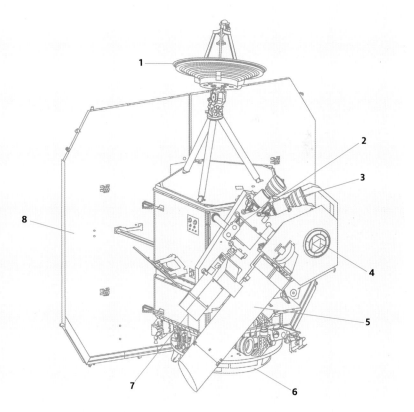

On January 12, 2005, a Delta II rocket launched the Deep Impact probe into space on a five-month trip to Comet 9P/Tempel 1, at a speed of 103,000 kilometers per hour. The spacecraft consisted of two elements: a 350-kilogram impactor, held in place by a copper dome and equipped with a close-up camera, and a space platform serving as a bus, equipped with two high-performance cameras and an infrared spectrometer. The mission's telescopic digital camera can produce images with 2 meters/pixel resolution, and the wide-angle camera, 10 meters/pixel.

As the spacecraft approached the comet on July 3, 2005, the impactor was released and fired its engines toward the target. Using its independent navigation system and camera, the impactor adjusted its trajectory and struck the comet nucleus 24 hours later, on July 4, with a relative velocity of 37,000 kilometers per hour. The collision released energy equivalent to the explosion of 5 tons of TNT, leading to a six-fold increase in the comet's brightness. As the probe flew past the comet a few minutes later, it transmitted nearly 4,500 images, but due to the huge, fine-powder dust cloud caused by the impactor, it could not image the resultant crater.

STEP-BY-STEP SEQUENCE OF THE COLLISION BETWEEN THE IMPACTOR AND COMET TEMPEL 1

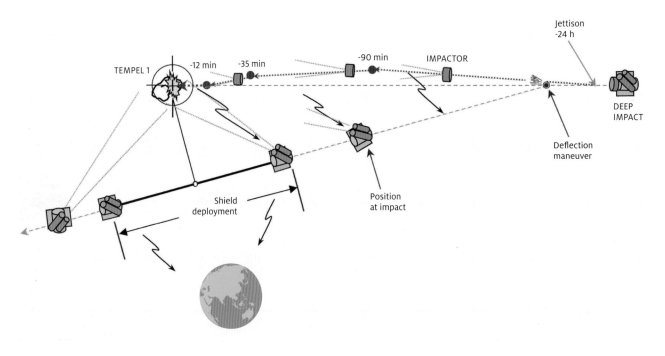

The Rosetta probe (see page 313), located some 80 million kilometers away, was able to observe the cloud of gas and dust after the impact, as were the Hubble Space Telescope and several Earthbound telescopes, both professional and amateur. The spectroscopic analyses of the impact cloud revealed the presence of ice, as well as clay silicates and carbonates.

Due to its spectacular nature, this phase of the Deep Impact Mission was well covered by the media. More than 10,000 people were able to watch the impact "live" on a giant screen at Waikiki Beach in Hawaii on Independence Day. Mathematical models had predicted that a 50- to 100-meter diameter crater would form on the comet. It was later estimated that the impact had excavated a 100-meter wide and 30-meter deep crater, spewing 250,000 tons of ice and 15,000 tons of dust into space. Despite these apparently gigantic quantities, the impact can be compared to a mosquito bite on an elephant's skin. It has been estimated that the shock of the impact resulted in a deviation of at most 10 centimeters in the comet's trajectory.

NASA subsequently approved an extension of the mission. Under the name of EPOXI (Extrasolar Planet Observation and Deep Impact Extended Investigation), the probe flew past Comet Hartley 2 at a distance of less than 700 kilometers in November 2010. Continuing on its voyage, its telescope will attempt to observe transits of exoplanets, known to orbit some nearby stars. For its part, the Stardust probe (see page 309) once again visited Comet 9P/Tempel 1 in 2011, to fully document the effects of the Deep Impact collision under more favorable viewing conditions.

ABOVE

This image was transmitted 67 seconds after the impactor struck the nucleus of Comet Tempel 1 at 37,000 kilometers per hour. The energy liberated was equivalent to that of 5 tons of TNT. The bright light is due to the Sun's reflection off the ejected material. The latter, composed mainly of fine dust, formed a dense cloud that masked the view of the crater from Deep Impact's cameras during the flyby. The collision created a crater 100 meters in diameter.

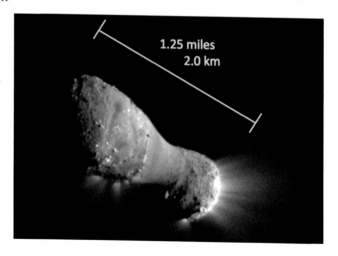

RIGHT

This medium-resolution image of the small nucleus of Comet Hartley 2 was captured by the EPOXI mission in November 2010. Many dust and gas jets are visible emanating from the comet's sunlit side.

Dwarf Planets and Asteroids

Pluto and dwarf planets

At the end of the 19th century, astronomers became aware of the small perturbations in the orbit of Neptune. These anomalies had to be caused by a more remote body, dubbed "Planet X" by Percival Lowell, who looked for it relentlessly from his observatory in Flagstaff, Arizona, until his death in 1916. Thanks to the acquisition of a new machine, a blink comparator, it now became possible to quickly scan pairs of photographic plates that had been taken a few days apart. It was this way that a young intern, Clyde W. Tombaugh, discovered a ninth planet on February 18, 1930. It would be called Pluto after both the god of the underworld and for Percival Lowell's initials. PL became its astronomical symbol. Although the search for Planet X was now over, the discovery of other trans-Neptunian bodies, including Erin in 2005 (which is larger and more massive than Pluto), as well as Makemake, Orcus and Sedna, has reopened the debate as to what such bodies really are. On August 24, 2006, the International Astronomical Union (IAU), amid considerable controversy, decided to classify Pluto as a dwarf planet.

Clyde W. Tombaugh (1906–1997) is shown here during his internship at Lowell Observatory in Flagstaff, Arizona.

OPPOSITE PAGE
Artist's impression of the double asteroid Antiope.

PLUTO DATA

Equatorial diameter: **2,306 km**
Mass: **0.002** (Earth = 1)
Mean distance from the Sun: **39.53 AU** (Earth = 1)
Rotation period: **6.39 days**
Orbital period: **248 years**
Mean ground temperature: **-229 °C**
Surface gravity: **0.06 g**
Escape velocity: **1.27 km/s**
Moons: **3 (Charon, Nix, Hydra)**

DISCOVERY OF THE PLANET PLUTO

After having examined thousands of paired photographic plates taken a few days apart with a blink comparator, Tombaugh discovered a new planet, Pluto, in 1930.

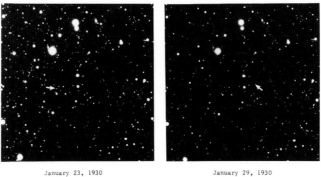

January 23, 1930   January 29, 1930

Pluto's diameter and mass are difficult parameters to measure. The latest estimates show that this planet is smaller in diameter (2,306 kilometers) and less massive than the Moon and Jupiter's four Galilean satellites. Pluto and its largest satellite Charon, named after the underworld's gatekeeper in Greek mythology, constitute a double planetary system, accompanied by the smaller moons Nix and Hydra, discovered in 2005.

BELOW

The relative sizes of the largest trans-Neptunian bodies known.

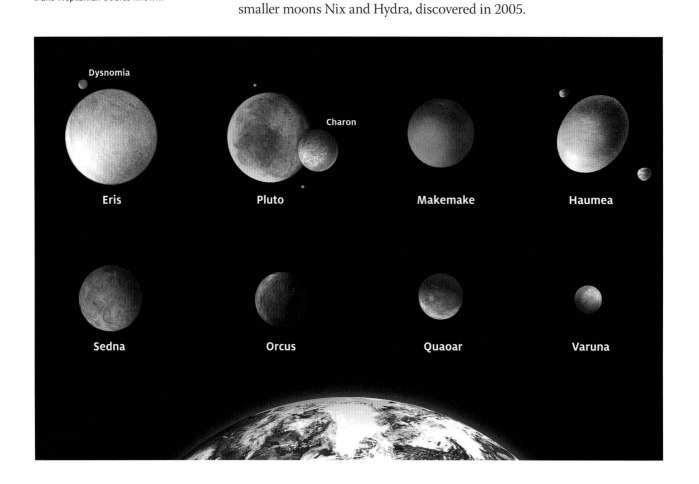

Dysnomia

Eris

Charon

Pluto

Makemake

Haumea

Sedna

Orcus

Quaoar

Varuna

Pluto will be the last planet in our solar system to be visited by a space probe when New Horizons flies by it in 2015 (see page 335).

The dwarf planet Sedna, with a diameter estimated between 1,200 and 1,800 kilometers, is a non-cometary object with the most eccentric elliptical orbit known: it can recede up to 897 AU from the Sun (see diagram on page 365).

Eris, named after the Greek goddess of discord, is the largest of the dwarf planets in the solar system, measuring between 2,400 and 3,000 kilometers in diameter, which qualifies it as the tenth planet to be discovered. Eris and its moon, Dysnomia, are at 97 AU from the Sun, three times the distance of Pluto and beyond the Kuiper Belt. ⊵

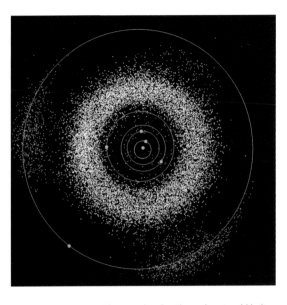

Diagram showing the main asteroid belt (in white) and the Trojan asteroids with the orbit of Jupiter (in green).

The Asteroids

The main asteroid belt, consisting of millions of small, rocky and metallic bodies, migrates between the orbits of Mars and Jupiter. The name asteroid means "star-like object." However, they are probably the remnants of a rocky protoplanet that did not have enough mass to accrete as the solar system was forming, and belong to a group of bodies termed minor planets. Ceres, the first asteroid to be discovered (in 1801), by itself accounts for 25 percent of the total of the asteroid belt. There are only a dozen asteroids with diameters exceeding 150 kilometers. The two small moons of Mars, Phobos and Deimos, are probably asteroids captured by the gravitational field of the red planet.

In astronomical terminology, the term meteor is used when referring to the flashing "shooting star" in the sky, and a meteorite is the actual solid body that survives the burn-up and lands on the ground. They represent solar system debris from comets, asteroids and other sources. More than 2,000 meteorites are found on Earth almost every day. In the past, however, at least one large meteorite had a major impact on the evolution of life on Earth. According to a theory proposed by Louis and Walter Alvarez, the rapid mass extinction 65 million years ago of 50 percent of all plant and animal species, most notably the dinosaurs, followed the impact of a meteorite at least 10 kilometers in diameter. The immense Chicxulub crater in the Gulf of Mexico dates from that time. This impact also defines the transition between the Cretaceous and Tertiary periods, evidenced by a sedimentary layer rich in iridium, a metal characteristic of meteorites.

This major event in the evolution of life on our planet most likely opened the door for the rise of mammals and birds. Several large Near-Earth Asteroids (NEAs: 900 have been detected to date), follow trajectories that could bring them close enough to our planet that they are potentially dangerous. Although the risks of such collisions are nearly zero over the coming several thousand years, the international community has nonetheless taken precautionary measures to monitor NEAs through a cooperative surveillance system.

Asteroids are a unique source of information on the birth and evolution of the solar system. The goal of the current Dawn space mission is to first orbit the asteroid Vesta and then Ceres between 2011 and 2015 (see page 340).

Willamette, the largest meteorite discovered in the United States, weighs more than 15 tons. It is currently on display at the New York Museum of Natural History.

Arizona's famous Meteor Crater, also known as the Barringer Crater, in honor of the American engineer Daniel Barringer (1860–1929), who first suggested it was the result of a giant meteorite that fell there about 50,000 years ago.

Galileo

THE GALILEO MISSION to Jupiter followed a trajectory that required several successive gravitational assists, provided by flybys of Venus and Earth. This complex trajectory brought Galileo into the asteroid belt twice before moving on to Jupiter. These two entries provided the first opportunity to closely observe some asteroids. On October 29, 1991, the probe came to within 1,600 kilometers of 951 Gaspra and took several images and spectroscopic measurements of this irregularly shaped asteroid, before leaving it at a speed of 8 kilometers per second (28,800 kilometers per hour). On August 28, 1993, 22 months after the short encounter with 951 Gaspra, Galileo cruised past the asteroid 243 Ida, from a distance of 2,400 kilometers. Images of Ida provided an unexpected surprise; the asteroid was accompanied by a tiny moon, dubbed Dactyl. Only 1.4 kilometers in diameter, Dactyl was the first natural satellite of an asteroid to be discovered. Spectrometric data obtained by Galileo showed that Ida and Dactyl have slightly different mineral composition. Nevertheless, most experts think that the two are the products of a larger object that disintegrated more than a billion years before. That remains a hypothesis, however. ✤

The Galileo space mission was the first to include an encounter with an asteroid: Gaspra is an irregularly shaped asteroid-belt object, about 19 kilometers long.

Galileo also encountered Ida (56 kilometers long), another asteroid-belt object, which was discovered in 1884.

LEFT
The tiny moon Dactyl, which orbits around Ida.

NEAR Shoemaker

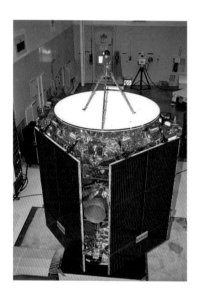

The NEAR spacecraft was octagonal in shape. It was designed and built at the Applied Physics Lab at Johns Hopkins University in Baltimore.

THE NEAR (Near-Earth Asteroid Rendezvous) mission was part of NASA's Discovery program, intended to develop highly targeted, low-cost missions (less than US$150 million), with maximum deployment times of 36 months between approval and launch. The main objective of the NEAR mission was to study the physical characteristics of the asteroid Eros from close orbit over a period of one year. Eros was the first NEA (Near-Earth Asteroid) to be discovered, in 1889. Eros is an S-type or stony asteroid, with an elongated, irregular shape (34.4 x 11.2 x 11.2 kilometers) and is known to cross the orbit of Mars. It is thought to be more massive than the asteroid that caused the Chicxulub crater in Yucatan. Since it is not in a stable orbit, Eros could cross the Earth's orbit in a few million years.

The NEAR spacecraft was equipped with the following instruments for its mission: a multi-spectral digital camera, an X-ray and gamma-ray spectrometer, a magnetometer and an infrared imaging spectrometer. Following its launch by a Delta 7925-8 rocket on February 17, 1996, NEAR was placed into sleeper mode while cruising to its target. It was reactivated a few days before scheduled to fly by the asteroid 253 Matilda on June 27, 1997, at a distance of 1,200 kilometers. Some 61 kilometers in diameter, Matilda is a C-class (carbon) asteroid, and one of the largest visited by a space probe at that time. One of its distinguishing features is a very slow rotation period, taking 17.4 days to turn on its axis. On July 3, 1997, NEAR made a trajectory adjustment by igniting its engines to slow down and lower its perihelion. On January 23, 1998, the probe got a gravitational boost from a 540-kilometer flyby of Earth, changing its orbital inclination from 0.5 degrees to 10.2 degrees, to bring it in line with the orbit of Eros.

This mosaic of images of the asteroid Matilda was taken by NEAR from a distance of 2,400 kilometers and shows a 10-kilometer deep crater at its center.

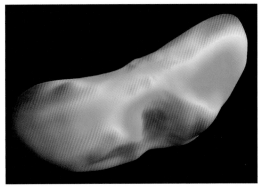

Eros is the largest of the NEA (Near-Earth Asteroids) group.

A topographic model of the asteroid Eros. Depressions are shown in blue and elevations in red.

MAIN COMPONENTS OF NEAR SHOEMAKER

1. High-gain antenna
2. Solar panel
3. Magnetometer
4. X-ray detector
5. Gamma-ray detector
6. Laser altimeters
7. Multi-spectral imager
8. Near-infrared spectrometer

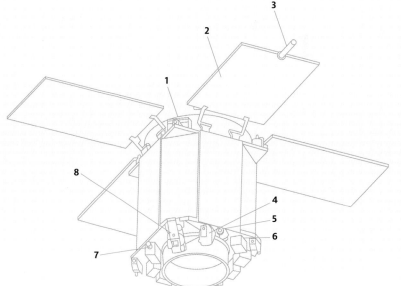

The first planned engine burn intended for the rendezvous did not take place because of a computer glitch, leading to loss of contact for more than 24 hours. As a result, NEAR would not be able to reach Eros until December 23, 1998, at a distance of 3,827 kilometers and moving at 965 meters per second. Mission control decided to place the probe into the same orbit as Eros and wait for a full rotation around the Sun. Thirteen months later, on February 14, 2000, NEAR entered a 321 x 366 kilometer orbit around Eros, first making sure Eros did not have a moon that might collide with the spacecraft. This placed the probe in position to explore and study all aspects of Eros from different angles for several months. Beginning on January 24, 2001, NEAR began a gradual approach of Eros until it was able to soft land on the surface, thereby becoming the first spacecraft to land on an asteroid. It remains there to this day. To the surprise of mission controllers, the probe stayed operational after this impact.

After receiving a final transmission from NEAR on February 28, 2001, NASA decided to end the mission. On March 14, 2000, a month after it entered orbit, NASA officially renamed the spacecraft NEAR Shoemaker, in honor of geologist Eugene M. Shoemaker (see following page).

NEAR's landing trajectory onto Eros, shown from its 35-kilometer high orbit above the asteroid.

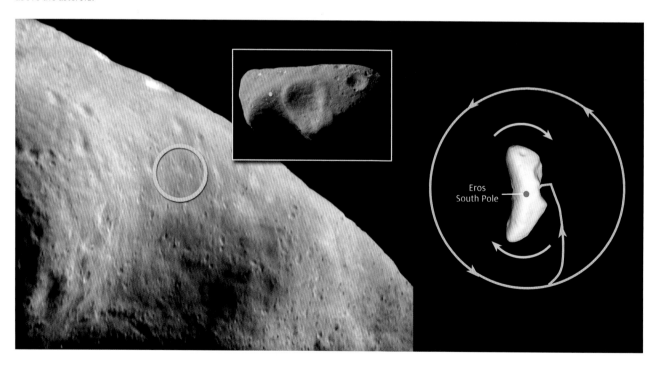

Eros
South Pole

EUGENE "GENE" M. SHOEMAKER

(1928–1997)

American geologist Eugene "Gene" M. Shoemaker was one of the founders of the science of planetology. It was after studying the Barringer crater in Arizona, as well as those resulting from underground nuclear tests, that he proposed that many such sudden geologic changes on Earth were due to impacts by asteroids. He also worked closely with NASA on the Ranger missions, which demonstrated that the lunar surface was covered with impact craters, and helped train the first astronauts. He is also well known as the co-discoverer, along with his wife and David Levy, of comet Shoemaker-Levy 9, which crashed into Jupiter in 1994 (see page 297). It was during his many expeditions looking for "hidden" impact craters that he lost his life, in a car accident in Australia. A commemorative capsule carrying some of his ashes on board the Lunar Prospector probe made Eugene M. Shoemaker the only person to be sent to rest on the Moon.

Hayabusa

The 510-kilogram MUSES-C/Hayabusa spacecraft being assembled at a JAXA facility. The atmospheric re-entry capsule, visible at right, returned to Earth in June 2010.

RIGHT
Artist's rendition of Hayabusa collecting a surface sample from the asteroid Itokawa.

HAYABUSA ("falcon" in Japanese), initially called MUSES-C, was a robotic space mission run by JAXA (Japan Aerospace Exploration Agency). Its primary mission was to collect and return to Earth a surface sample from an asteroid. The asteroid 1998 SF 36 was selected as the target: it crosses the orbits of both Earth and Mars and is a rocky body of moderate size (540 meters long). It was discovered by the LINEAR (Lincoln Near-Earth Asteroid Research) program, and renamed Itokawa, in honor of Hideo Itokawa (1912–1999), humanist and "father" of Japanese aeronautics. Hayabusa was launched on May 9, 2003, from the Kagoshima Space Center by an M-5 solid-fuel rocket. Hayabusa was equipped with an imaging and navigation digital camera, a laser altimeter and infrared and X-ray spectrometers.

A conical-shaped trough was added to transfer surface samples into a hermetically sealed chamber in the return capsule. The mission also carried a miniature surface rover called MINERVA (Micro/Nano Experimental Robot Vehicle for Asteroid), intended for the asteroid's surface.

The spacecraft was propelled by four ion-thrust engines to permit course adjustments for rendezvous with the asteroid. The engines were successfully test-fired from May 27 to mid-June, 2003. At the end of that year, a violent solar eruption damaged the probe's solar panels. As a result, its engines could no longer function at maximum capacity and the spacecraft slowed down. The rendezvous with the asteroid was consequently postponed to 2005. This unfortunate incident meant that Hayabusa's time near the asteroid was greatly shortened. Because of unavoidable orbital dynamics, it had to turn back toward the Earth in November of that year. The spacecraft slowed to a stationary position about 20 kilometers from Itokawa on September 12 and began its initial mapping of the asteroid, and moved to within 7 kilometers on October 2 for further observations. On November 4, the first landing effort took place, but it was aborted because of a failure in the probe's navigation system at 700 meters above the asteroid's surface.

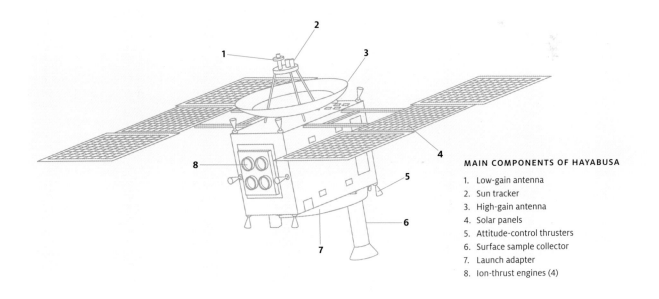

MAIN COMPONENTS OF HAYABUSA

1. Low-gain antenna
2. Sun tracker
3. High-gain antenna
4. Solar panels
5. Attitude-control thrusters
6. Surface sample collector
7. Launch adapter
8. Ion-thrust engines (4)

No surface impact craters are evident on the asteroid Itokawa. Its surface is composed of different sizes of rock fragments. The white rectangle shows Hayabusa's landing point and the International Space Station is shown as a scale reference.

A second attempt on November 12 also did not go as planned. The little lander MINERVA was released too late and too high at 55 meters from the surface, just as the probe began to automatically distance itself from the asteroid. MINERVA was lost in space. Hayabusa did manage to touch down on the surface of Itokawa and rest after several bounces, but did not obtain a surface sample. A second landing took place on November 25. This time, two projectiles were released in an effort to stir up a cloud of dust that could be sampled, but that effort remained inconclusive. After 30 minutes on the asteroid's surface the spacecraft took off again. Unfortunately, efforts to collect surface samples did not work, but Hayabusa was still the first spacecraft to land on and return from a celestial object other than the Moon. Because the small asteroid Itokawa has such low gravitational force, project managers were confident that a small amount of surface dust was collected when the probe was on the surface, and they sealed the return capsule. Radio contact with the spacecraft was interrupted on December 9, 2005, and could not be re-established until March of 2006. The ion thrusters were working normally, thereby assuring that Hayabusa was on the right trajectory toward the Earth. The return capsule was released from Hayabusa 300,000 kilometers from Earth in order to assume a ballistic re-entry trajectory. On June 13, 2010, it entered the Earth's atmosphere over southern Australia like a fireball. The capsule was designed to withstand a harsh decceleration of 25 g and its heat shield protected it against temperatures 30 times greater than an Apollo capsule was exposed to. The parachute opened as planned and the capsule landed safely in the Woomera Protected Area. Subsequent analysis by JAXA scientists of the micro sample particles, most less than 10 microns in size, showed them to be of extraterrestrial origin, namely from the asteroid Itokawa. Ultimately, despite all the problems, the Hayabusa Mission was the first to successfully retrieve samples from an astronomical body farther away than the Moon.

New Horizons

Nasa's robotic mission to pluto, New Horizons, is currently underway. It will mark the first time that a space probe visits the dwarf planet and its three moons, Charon, Nix and Hydra. The mission's principal investigator, S. Alan Stern, confirmed that the ashes of Clyde W. Tombaugh, who discovered Pluto, were placed on board in his memory.

New Horizon's encounter with Pluto is predicted for July 14, 2015. The second phase of the mission is to continue toward the Kuiper Belt, where it will encounter several interesting, but so far unidentified, trans-Neptunian objects. New Horizons was launched from Cape Canaveral on January 19, 2006, on a direct course to the outer solar system, by Lockheed Martin's powerful Atlas V-551 rocket, assisted with a Boeing third-stage STAR 48B booster rocket.

The team of scientists and technicians of the New Horizons mission. Dr. Alan Stern, of the Southwest Research Institute and the principal investigator, is seated front center.

With an escape velocity of 16 kilometers per second, the spacecraft was the fastest ever to leave Earth. It arrived close to Jupiter on February 28, 2007 (see page 254), where it received a gravitational boost, and then crossed the orbit of Saturn on June 8, 2008. At more than 30 AU, Pluto is too far from the Sun to adequately power New Horizons via solar panels of any reasonable size. Instead, the craft is powered by a thermoelectric nuclear reactor. To limit costs, a backup nuclear generator from the Cassini-Huygens mission (see page 238) was used. The unit contains 11 kilograms of plutonium 238 and will furnish 240 watts of electricity at 30 volts during the Pluto flyby. Telecommunications between New Horizons and the Deep Space Network will be carried out at a modest 1 kilobyte per second rate at the distance of Pluto, which is very slow relative to the 38 kilobytes per second from Jupiter. The high-gain antenna is two meters in diameter, affixed on top of the triangular New Horizons platform, giving it the odd appearance of a grand piano tagged with a parabolic antenna. The probe is equipped with 16 thrusters for trajectory corrections and attitude control, which also depends on optical star-tracker cameras.

BELOW, LEFT
Installation of the nuclear thermoelectric generator on the New Horizons bus.

BELOW, RIGHT
The first mission to an unexplored planet since Voyager, New Horizons was launched toward Pluto on January 19, 2006, on board an Atlas V rocket.

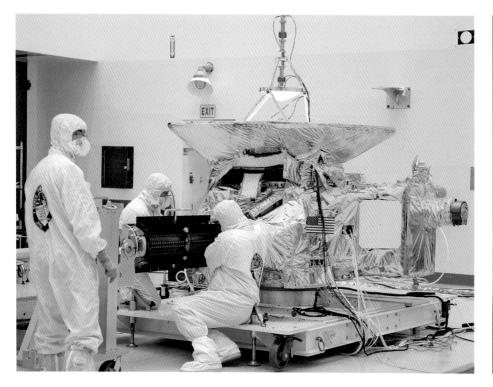

The New Horizons computer consists of two parallel information systems, one dedicated to mission command and science data management, the other to navigation and attitude control. Each system has built-in redundancy and is equipped with Mongoose-V processors, hardened versions of MIPS RISC R3000 clocked at 12 MHz. The spacecraft carries seven science instruments, including LORRI (Long Range Reconnaissance Imager), a high-resolution telescopic CCD imager for visible-light observations of planetary systems and other celestial bodies. The 21-centimeter diameter mirror telescope is made of silicon carbide, specifically chosen because of its mechanical and thermal stability. New Horizons' PERSI module (Pluto Exploration Remote Sensing Investigation) includes three instruments: a 6-centimeter aperture telescope and two high-resolution imaging spectrometers that cover the near-infrared to ultraviolet range. The UV spectrometer, an improved version of the one on board Rosetta (see page 313), will be particularly useful for chemical analysis of Pluto's thin atmosphere when the planet occults a background star.

MAIN COMPONENTS OF NEW HORIZONS

1. High-gain antenna
2. Solar-wind and plasma spectrometer
3. Imaging telescope
4. Star tracker
5. Visible and infrared imaging spectrometer
6. UV imaging spectrometer
7. Nuclear thermoelectric generator

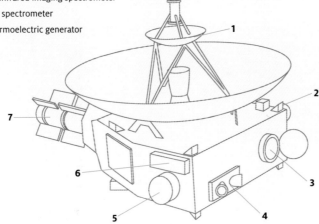

Artist's impression of New Horizons, with Pluto and its three moons in the background.

Two additional spectrometers, SWAP (Solar Wind Around Pluto) and PEPSSI (Pluto Energetic Particle Spectrometer Science Investigation), will measure interactions of various solar-wind particles with Pluto's atmosphere. The passive radiometer, REX (Radio Science Experiment), is equipped with an ultrastable oscillator, used to measure occultations and the Doppler effect of radio signals sent to Earth by the spacecraft. In addition, the probe carries the cosmic dust counter, VBSDC (Venetia Burney Student Dust Counter), which was built by students at the University of Colorado, Boulder in honor of Venetia Burney (1919–2009), the English girl who in 1930 suggested the name Pluto to Clyde W. Tombaugh for his new planet.

The first observations of Pluto are scheduled to begin about six months before the flyby, when the planet will only cover a few pixels of the CCD images. That should be enough to detect any new moons or rings around the planet and make trajectory adjustments to avoid possible collisions. Two months before the flyby, LORRI images of Pluto and Charon will be of higher resolution that those obtainable with the Hubble Space Telescope. Surface mapping of Pluto to a resolution of 40 kilometers should be possible a few days before flyby. During the actual flyby, from a distance of about 10,000 kilometers and a velocity of 11 kilometers per second, details down to 100 meters on Pluto should be visible. Due to its great distance from Earth and the limited communication capabilities, New Horizons will undertake the first-ever "uplink" (rather than downlink) radio science experiment. Thanks to powerful radio signals sent from the Earth, the REX radiometer will able to ascertain Pluto's mass and diameter with great precision. The first New Horizons images of Pluto and its moons will be sent in highly compressed formats in the days after the flyby, but it will take several months of processing before we will be able to fully appreciate their quality. After that, although the spacecraft will be nearly out of fuel, it may be able to reach one of the larger Kuiper Belt objects and observe it. Once past 55 AU, however, its radio signals will be too weak to transmit more data, since it will no longer have enough electric power to function normally.

In June 2011, New Horizons, moving at a speed of about 16 kilometers per second, will have exceeded a distance of 20 AU (about 3 billion kilometers) from Earth and its radio signal will take almost three hours to reach us. When the probe reaches Pluto in July of 2015, it will be some 31 AU (or 4.6 billion kilometers) from Earth.

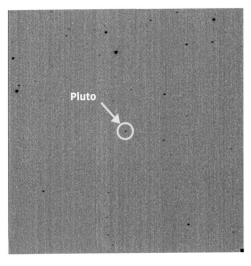

The first image of Pluto obtained by the high-resolution telescopic camera LORRI was taken on October 6, 2007, when New Horizons was still 3.6 billion kilometers from the dwarf planet!

Dawn

D AWN IS PART OF NASA'S DISCOVERY PROGRAM and currently on its way to its first destination. Its goal is to visit two principal bodies of the asteroid belt: Vesta in 2011–2012 and the dwarf planet Ceres in 2015. This marks the first attempt to orbit one celestial body and then move on to orbit another destination. To accomplish this, Dawn will be powered by three xenon ion-thrust engines, first used on Deep Space 1 (see page 306). The mission's name is based on the fact that it will explore Vesta and Ceres, two bodies dating back to the time of the formation of the solar system, when the Sun's rays were still scattered by the dust of the protoplanetary nebula.

Inspection of the complex tubular array of Dawn's ion-thrust propulsion system.

The 1,224-kilogram probe was launched from Cape Canaveral by a Delta 7925-H rocket on September 27, 2007, with an initial velocity of 11.2 kilometers per second. The xenon ion-thrusters added an additional 10 kilometers per second, setting a new speed record for a space mission. Its fuel reserves are 425 kilograms of xenon, and Dawn will use 275 to reach Vesta and another 110 to visit Ceres. The spacecraft received a gravitational assist on February 17, 2009, when it bypassed Mars. During its 550-kilometer flyby of the red planet, the probe calibrated its main camera, but due to a computer navigation error, which forced it into sleeper mode for safety, it was unable to calibrate its other instruments.

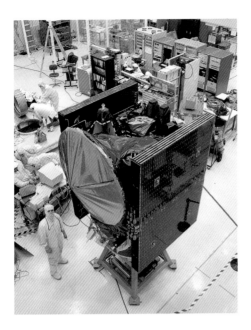

Dawn undergoing final inspection in the white room of the Orbital Sciences Corporation in Dulles, Virginia.

On the morning of September 27, 2007, Dawn is launched towards Mars and the asteroid belt by a Delta II Heavy rocket, which follows a parabolic trajectory.

PRINCIPAL COMPONENTS OF DAWN

1. Star tracker
2. Laser altimeter
3. Magnetometer support arm in extended configuration
4. High-gain antenna
5. Medium-gain antenna
6. Ion thrust engine
7. Hydrazine attitude-control thruster
8. Launch adapter
9. Solar panel
10. Radiator
11. Neutron and gamma-ray spectrometer
12. Imagers
13. Mapping spectrometer

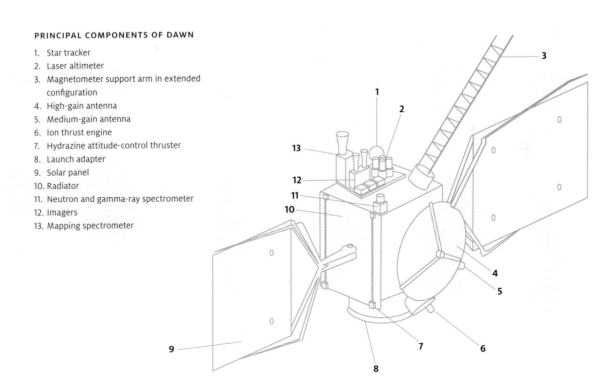

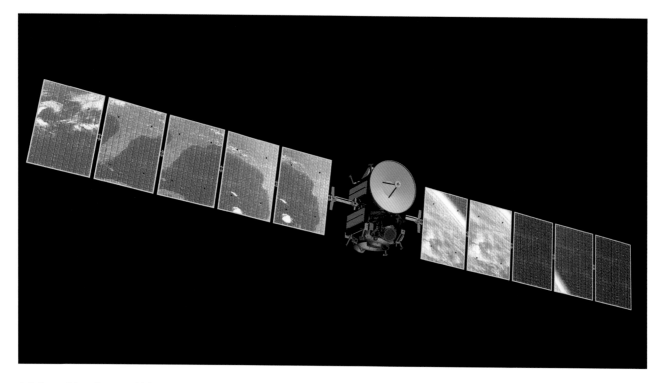

Artist's rendition of Dawn, with its extended solar panels reflecting an image of the Earth.

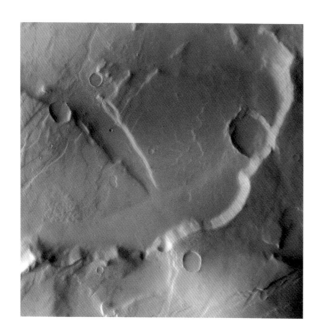

While passing Mars during a gravitational assist, Dawn was able to calibrate its camera.

Dawn is also equipped with a near-infrared spectrometer and a neutron and gamma-ray spectrometer, two instruments designed to map the mineral composition of the surfaces of Vesta and Ceres.

These two asteroids were selected for study because they are very different in structure. Vesta's surface is typical of basalt-rich and water-poor asteroids, while Ceres is composed of water-rich carbonaceous chondrite material. Having undergone multiple collisions, Vesta is known to be the source of numerous smaller asteroids (Vestoids), which account for 5 percent of all meteorites found on Earth.

Radioisotopic dating of these meteorites shows that Vesta underwent significant differentiation, due to fusion and solidification, three million years ago, a very short time by astronomical standards. Dawn is scheduled to arrive near Vesta in July 2011, and after six months of observation, will leave for Ceres in April 2012. It will reach Ceres in February 2015 and the mission will end in July 2015.

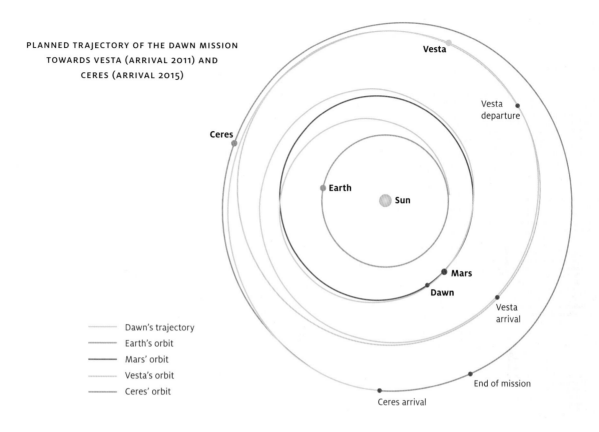

PLANNED TRAJECTORY OF THE DAWN MISSION
TOWARDS VESTA (ARRIVAL 2011) AND
CERES (ARRIVAL 2015)

Vesta

Vesta
departure

Ceres

Earth

Sun

Mars

Dawn

Vesta
arrival

End of mission

Ceres arrival

—— Dawn's trajectory
—— Earth's orbit
—— Mars' orbit
—— Vesta's orbit
—— Ceres' orbit

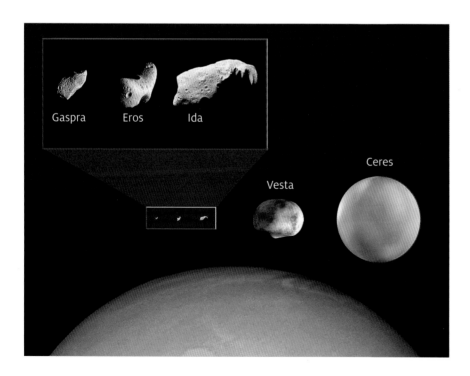

Gaspra   Eros   Ida

Vesta

Ceres

Relative sizes of the dwarf planet Ceres
and the asteroid Vesta, which will be
visited during the Dawn mission, with
Mars shown in the background. The
relative sizes of asteroids Gaspra, Eros
and Ida are shown in the small rectangle.

Future Approved Missions

THE ENTHUSIASM AND CURIOSITY of the worldwide astronautical community know no limits. Consequently, numerous proposals for exploratory missions are submitted each year to the various space agencies. Since final selection of such proposals is based on their scientific and technological merits, as well as on budgetary and political considerations, it is impossible to predict with any certainty which have the best chance of being approved. For that reason, we will describe here only those that have already been given the green light.

2011

The Russian interplanetary mission Phobos-Grunt (below), which was conceived and developed by the company NPO Lavochkin, is nothing if not ambitious. Plans are to send a lander to the Martian moon Phobos, collect a soil sample and return it to Earth via an atmospheric return capsule. Originally approved for 2009, the Chinese Mars orbiter Yinghuo-1 will be part of the joint mission, scheduled for launch from the Baikonur Cosmodrome by a Zenit-Fregat rocket in November 2011.

The small, Chinese-built Yinghuo probe (a play on words, referring jointly to "Mars" and "firefly") will orbit Mars to map the surface and study its atmosphere. Since the Chinese do not have the capability yet, Yinghuo-1 communications and control will be carried out via an array of European and Russian radiotelescopes.

OPPOSITE PAGE

This massive aeroshell will protect the Mars Science Laboratory lander during the descent through the Martian atmosphere.

The Juno Mission to Jupiter is part of NASA's New Frontiers Program, which focuses on missions with high science returns potential but moderate costs (around US$700 million). After a five-year voyage, the spacecraft Juno (above) will be placed in a polar orbit around the giant planet in 2016, in order to investigate Jupiter's internal structure (below), its atmosphere and magnetosphere. The probe's color camera will also provide details on Jupiter's polar regions. Because of significant advances in solar-panel technology since the Galileo mission, Juno will be able to generate electricity from sunlight rather than nuclear power.

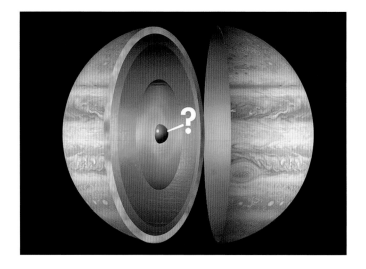

NASA's Mars Science Laboratory rover, named Curiosity (below), is scheduled for launch at the end of 2011 by an Atlas V rocket. It will attempt the first-ever precision landing on the red planet. Three times heavier than Spirit or Opportunity, it will be equipped with far more sophisticated instrumentation than all the preceding missions to Mars. Its principal objective will be to search for evidence of life in the Martian soil, be it fossilized or extant. Because the rover is nuclear powered, it can operate under all conditions and latitudes, as well as day or night. Designed to function autonomously for at least one Martian year, the complex rover will cost an estimated US$2 billion.

2013

Buoyed by its recent success with Chandrayaan-1, the Indian Space Research Organization (ISRO) intends to launch a second lunar exploration probe, Chandrayaan-2, in collaboration with the Russian Federal Space Agency (RKA). This mission will combine an orbiter and a lander that will deploy an RKA-built rover to explore the Moon's polar regions.

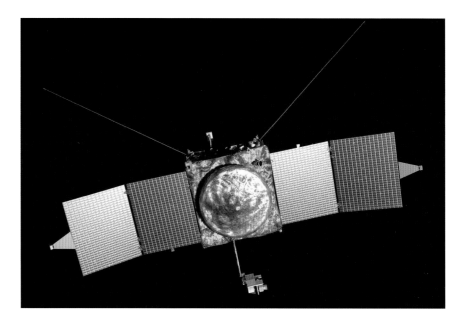

NASA plans to launch a mission to Mars called MAVEN (Mars Atmosphere and Volatile EvolutioN), as part of its Mars Scout program. The launch is scheduled for the end of 2013 and will orbit Mars in 2014. MAVEN (above) will study the ionic composition of the Martian upper atmosphere and its interaction with the solar wind.

2014

A collaborative effort between ESA and JAXA, the BepiColombo mission (left), is named in honor of the Italian scientist Giuseppe (Bepi) Colombo (1920–1984), whose brilliant work on the orbital dynamics of Mercury contributed to the success of Mariner 10 (see page 101). Although the spacecraft's configuration has not yet been finalized, it will consist of two orbiters: MPO (Mercury Planet Orbiter), built by ESA, and MMO (Mercury Magnetospheric Orbiter), built by JAXA. Particular emphasis will be placed on thermal protection for the two orbiters, which will have to withstand temperatures as high as 350°C. BepiColombo will be launched in 2014 by a Soyuz-Fregat rocket from the Kourou Space Center and arrive at its destination in 2020.

2016

On May 20, 2010, the Japanese space agency JAXA launched the Venus orbiter Akatsuki (also known as Venus Climate Orbiter or Planet-C) atop an H-IIA 202 rocket from its launch site at Tanegashima for a planned two-year mission. Equipped with several infrared cameras and other imagers, the probe (below) will attempt to verify that an active volcano and thunderstorms are present on Venus. Wind speeds around 300 kilometers per hour have been measured at the Venusian cloud tops. This is 60 times the speed of the planet's rotation, compared to Earth where the fastest winds are only 10 to 20 percent of the rotation speed. Known as super rotation, this phenomenon is one of the mysteries the spacecraft will attempt to clarify. By entering a highly elliptical orbit, 300 by 60,000 kilometers, the probe will alternate close views with distant ones, in an effort to better understand the meteorology of our sister planet. Unfortunately, an orbital insertion attempt on December 7, 2010, did not succeed, due to an engine problem. As a result, Akatsuki assumed a heliocentric trajectory that will bring it back to Venus in 2016 for another attempt at orbiting the planet.

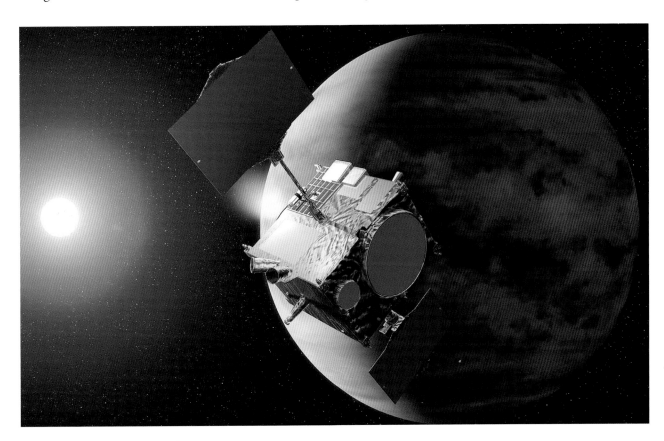

2017

The principal objective of the ESA's Solar Orbiter mission (below) will be to obtain new data on the activity of the Sun's heliosphere at its polar regions, which have not been explored in much detail to date. NASA will supply the Atlas V rocket and launch logistics for the mission, scheduled for deployment in January of 2017. Solar Orbiter will be able to approach the Sun as close as 0.22 AU, that is to say, well within the orbit of Mercury.

2018

NASA's Solar Probe Plus (opposite page), will attempt to answer two mysteries about the Sun: why is the temperature of the corona so much higher than that of the photosphere and what drives the solar wind? By gradually nearing the Sun through successively closer orbits, the probe will actually enter the outer fringes of the corona and sample it directly. This is something no other space mission has dared to do.

ExoMars, currently in preparation, is the first mission of the Aurora program, the ESA's extensive planetary exploration agenda. The mission's goal is to land a six-wheel rover (below), similar in size to the American rovers Spirit and Opportunity, and like them, powered by solar energy. The orbiting space bus will release the descent module containing the rover after circling the planet for several months, or until the dust and sandstorm season is over. The rover will be equipped with a science package designed to look for signatures of Martian life and to study the planet's geology. ExoMars and Mars Science Laboratory will be precursors to a joint ESA and NASA effort scheduled between 2020 and 2025, to collect samples of Martian soil and return them to Earth.

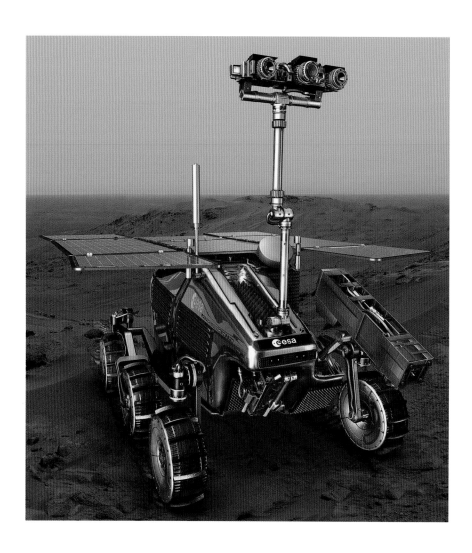

2020

Some of the most pressing questions in planetary astronomy center around two of Jupiter's moons: the possibility that life may exist in the ocean of the ice-covered Europa, and the nature of Ganymede's magnetic field and whether an ocean exists beneath its surface. That is why NASA gave priority in 2009 to an international mission, EJSM (Europa Jupiter System Mission), in collaboration with ESA, to study the Jovian moons, instead of an exploratory mission to Saturn's moon Titan. NASA will have primary responsibility for the JEO, Jupiter Europa Orbiter (top left), which, after several flybys of the moon Io, will study Europa's ocean in detail. For its part, the ESA will construct the orbital probe JGO, Jupiter Ganymede Orbiter (top right), which will make 19 flybys of the moon Callisto before entering orbit around Ganymede. The spacecraft will be launched separately in 2020 along an indirect VEEGA trajectory (Venus-Earth-Earth Gravity Assist), but will arrive at their destinations together in 2026. The two orbiters will study the Jovian system until at least 2029.

Looking Ahead

THESE FIRST 50 YEARS of solar system exploration (1959–2009) profoundly altered our understanding of the Earth and the Universe. There is no doubt that space probes and space telescopes have played dominant roles in ushering in this new era. Despite this, the debate between proponents of crewed space flight and those favoring robotic missions continues. We shall avoid entering this controversy, but instead express our sincere hope that a balance will be found between the two approaches in the future. This would help to avoid the setbacks caused by the American space program's dependence on the space shuttle. Although the space shuttle program has been in operation since 1981, and it is certainly a great technological achievement, it has never fully met the expectations of its users, primarily because it can only operate in low orbit and has proven more problematic than originally anticipated. Although the shuttle program has been risky and has cost the lives of two teams of astronauts, it has nonetheless fulfilled its primary mission, namely the construction and support of the International Space Station (ISS). The ISS is an indispensable laboratory and training vessel for extended Earth orbits and for preparing human beings for long-term space missions. The Vision for Space Exploration program, announced by U.S. President George W. Bush as the new NASA agenda on January 14, 2004, outlined this clearly. It was to continue with robotic missions but culminate with a manned mission to Mars in 2030. However, President Barak Obama outlined a different, more flexible agenda for NASA on April 15, 2010. Called "Flexible Path," this program emphasized robotic exploration of the inner solar system, with spacecraft orbiting at the Lagrangian points and crewed flybys of the planets Venus and Mars, in preparation for future human exploration of Mars.

Since the Luna 1 mission in 1959, all nations involved in space exploration have acquired considerable expertise and learned from their many historic successes as well as their failures. Over this period, the design of space probes has greatly benefited from unprecedented advances in electronics, information technology and telecommunications. Looking ahead, we can foresee what kind of technological advances will play key roles in future spacecraft design.

OPPOSITE PAGE

A short trip from Earth, returning to the Moon will help us prepare for a crewed mission to Mars.

Miniaturization

Since weight is a critical factor in launching interplanetary probes, miniaturization of cameras, scientific instrumentations and other subsystems will make it possible to make lighter spacecraft, which can be sent into more direct trajectories and will navigate and function with greater independence. For example, in 2003 the panoply of scientific instruments carried by the technologically advanced ESA mission SMART-1 only weighed 19 kilograms.

Modularity

Modular subsystems that have already proven successful on prior missions (imaging, telecommunications, power requirements, optical navigation and computers) become obvious choices for new spacecraft platforms. That also helps shorten the mission development time and lowers costs: the Venus Express mission is based on components previously used in the Mars Express and Rosetta spacecraft.

Internationalization

Space missions of high scientific value are becoming increasingly complex and costly. Consequently, cooperation among various space agencies is becoming more common, both in terms of development and cost-sharing. Space exploration will continue to be risky business in the foreseeable future, but international cooperation and information exchange increase the chances of success. A perfect example of this is the spectacular Cassini-Huygens mission to Saturn, which was jointly developed and managed by NASA and ESA and is still underway.

Autonomy

Interplanetary distances are simply too great to allow control of spacecraft trajectories in real time in order to help them avoid unforeseen hazards and obstacles in a hostile environment. For instance, even our close planetary neighbor, Mars, is far enough away that it takes a minimum of five minutes to communicate with it from Earth at the speed of light. Consequently, sending commands to a Mars probe and awaiting a response takes at least ten minutes. Since spacecraft computers have become more and more powerful, engineers are increasingly able to program them for autonomous operations. Thanks to software like Remote Agent, which was road tested during the Deep Space 1 mission (1998), space probes can now monitor their internal status and quickly take appropriate measures before communicating with Earth.

Similarly, operations such as atmospheric entries and landing, as well as roving on planetary surfaces, have benefited a great deal from advances in robotics, especially in the area of autonomous decision making. Consequently, the rover of the Mars Science Laboratory mission will be able to select its landing site based on local topographic data acquired moments before landing.

As the initial stages of solar-system exploration come to an end, our home planet is looking more and more like an oasis in the vast cosmic desert. The Kepler space telescope, launched on March 6, 2009, methodically surveys more than 100,000 stars in an area of sky between the constellations of Cygnus and Lyra for 3 years, seeking Earth-like planets. This will provide us with an indication of how many planetary oases might exist in our galaxy. The initial results point to a staggering number of planetary systems. Whether life might have evolved on any of these planets, let alone advanced civilizations, will be yet to be determined. Interstellar travel will not be possible until entirely new propulsion technologies have been developed. These would have to be hundred times more powerful than those currently used with our space probes, before we could even contemplate a visit to the closest stars.

In our solar system, the possibility that conditions conducive to life might be found on Mars, or in the oceans of Jupiter's moon Europa, has to be investigated thoroughly. Two exobiological missions to Mars (the first since the failure of Beagle 2 in 2003) are under development by NASA and ESA for deployment in this decade. The return to Earth of a sample of Martian soil is top priority now that the technology proposed seems realistic (pictured above). The discovery of life elsewhere in the solar system, or beyond, will be a defining moment in human history and will probably shift future space exploration in directions impossible to appreciate today.

"Tranquility base here, the Eagle has landed." Neil Armstrong took this photograph of Buzz Aldrin standing next to the lunar lander in Mare Tranquillitatis.

APPENDIX
THE APOLLO PROGRAM • 1963-1972

This historic space program has been exhaustively covered in many books and articles. We will content ourselves with summarizing its development, for the simple reason that vehicles used for crewed missions are not classified as space probes. Nevertheless, the challenge to the space race, which was taken up by President John F. Kennedy in 1961, was the definitive stimulus that led to several robotic lunar exploration missions. The Ranger, Surveyor and Lunar Orbiter probes contributed significantly to the Apollo program by playing the role of scouts to help locate the best landing sites.

To place things in the historical context of the era, it must be noted that the Soviet Union had reached a number of firsts in the conquest of space: launching the first artificial satellite, Sputnik-1 in 1957; sending the dog Laika in orbit the same year; launching the first space probe, Luna 1 in 1959; and sending Luna 2 to be the first human-made object to touch the Moon the same year. On April 12, 1961, Russian cosmonaut Yuri Gagarin became the first man to fly in space, and completed one full orbit around the Earth. The entire world perceived the Soviet Union's success in space as a measure of its technological and military supremacy. Deep international tension developed in April of 1961, when the United States suffered international humiliation with the Bay of Pigs incident and the failure of the Cuban invasion.

The race to the Moon began officially in 1961 when John F. Kennedy challenged the nation during his address to the United States Congress.

LEFT

The crew of the historic Apollo 11 mission (from left to right): Neil A. Armstrong, Michael Collins and Edwin E. "Buzz" Aldrin.

On July 16, 1969, the giant Saturn V rocket lifted off with the three astronauts of the Apollo 11 mission on board. Despite the power of the rocket's cryogenic engines, lifting the 3,040-ton mass took several seconds.

CONFIGURATION OF A CREWED APPOLO MISSION AT LAUNCH

1. Escape tower
2. Command Module
3. Service Module
4. Propulsion Engine Nozzle
5. Lunar Module
6. Third Stage of the Saturn V rocket

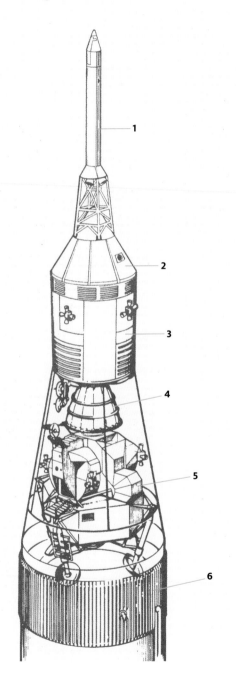

As a result, the American administration decided to set a national goal that would be both ambitious and unifying. On May 25, 1961, during an address to Congress, President John F. Kennedy proposed a challenge to race to the Moon: "I believe that this nation should commit itself to achieving the goal, before this decade is out, of landing a man on the moon and returning him safely to the Earth."

In 1963, he assigned this enormous responsibility to the newly formed agency NASA (National Aeronautics and Space Administration), and the Apollo program was born. Its success hinged on two entirely new heavy launchers, the Saturn IB and the Saturn V. These three-stage rockets with cryogenic engines were designed by Wernher Von Braun (see page 23) and his team at NASA's Marshall Space Flight Center. The Apollo program also benefited from advances made by the Mercury Space Program (1959–1963), which validated several technologies crucial for crewed space missions. The challenge was met with enthusiasm. Astronauts Neil Armstrong and Buzz Aldrin became the first humans to set foot on the Moon in July, 1969, before the deadline set by President Kennedy. After that, however, Americans quickly lost interest in lunar explorations by their astronauts. The last mission, Apollo 17, ended on December 19, 1972, with a drive in a lunar jeep and the collection of 110 kilograms of rocks to return to Earth for analysis. The Apollo program was the largest non-military technological endeavor ever undertaken by the United States. It cost more than US$150 billion (in 2011 dollars) and employed 400,000

This photograph of Buzz Aldrin's footprint was part of an experiment to test the mechanical properties of the lunar regolith (soil).

people in more than 20,000 companies and universities. In retrospect, we can conclude that the Soviets lost the race to the Moon primarily because they could not design as reliable and powerful a rocket as the Saturn V. More than forty years have passed since then and no other humans have had the opportunity to leave their footprints on the surface of the Moon. Recently, NASA announced an entirely new program, Flexible Path, which opens the possibility of a return to the Moon or a crewed mission to one of the Near-Earth Asteroids (NEAs) around 2020, as preludes to future expeditions to Mars.

The command and service modules of the Apollo mission.

Apollo 11's lunar module, the Eagle, on its return from the lunar surface, during the approach phase, before docking with the command module Columbia. Photographed by astronaut Michael Collins.

The maritime recovery operation of the Apollo 17 capsule and crew, on December 19, 1972. This was the last mission of this program.

Astronaut Harrison Schmitt, lunar module pilot and trained geologist, inspects a rock near the landing site in the Taurus-Littrow Valley, during an extra-vehicular activity of the Apollo 17 mission.

Key Steps in the Apollo Program

TEST FLIGHTS

AS-201 (February 26, 1966) First test of the Saturn IB rocket.

AS-203 "Apollo 3" (July 5, 1966) Weight test of fuel tank.

AS-202 "Apollo 2" (August 25, 1966) Suborbital test of the Saturn IB rocket and the command and service modules.

Apollo 4 (November 9, 1967) Test of the Saturn V rocket.

Apollo 5 (January 22 to February 12, 1968) Test of the Saturn IB rocket and the lunar module.

Apollo 6 (April 4, 1968) Test of the Saturn V rocket.

CREWED ORBITAL MISSIONS

Apollo 1 (January 27, 1967) Loss of the crew to fire during ground tests.
Crew: Virgil "Gus" Grissom, Edward White and Roger Chaffee

Apollo 7 (October 11–22, 1968) First crewed Apollo flight with three astronauts in orbit.
Crew: Walter M. Schirra, Donn Eisele and Walter Cunningham

Apollo 8 (December 21–27, 1968) First flight around the moon.
Crew: Frank Borman, Jim Lovell and William A. Anders

Apollo 9 (March 3–13, 1969) First crewed flight with the lunar lander module.
Crew: James McDivitt, David Scott and Russell Schweickart

Apollo 10 (May 18–26, 1969) First crewed flight around the moon with the lunar lander module.
Crew: Thomas Stafford, John W. Young and Eugene A. Cernan

CREWED MISSIONS WITH LUNAR LANDINGS

Apollo 11 (July 16–24, 1969) First human steps on the Moon.
Crew: Neil Armstrong, Michael Collins and Edwin "Buzz" Aldrin
Landing site: Mare Tranquillitatis (Sea of Tranquility)

Apollo 12 (November 14–24, 1969) First mission with a precise lunar landing.
Crew: Pete Conrad, Richard Gordon and Alan Bean.
Landing site: Oceanus Procellarum

Apollo 13 (April 11–17, 1970) Aborted mission following an explosion in the service module.
Crew: Jim Lovell, Jack Swigert and Fred Haise

Apollo 14 (January 31 to February 9, 1971) First round of lunar golf.
Crew: Alan Shepard, Stuart Roosa and Ed Mitchell
Landing site: Fra Mauro crater

Apollo 15 (July 26 to August 7, 1971) First mission with a lunar roving vehicle.
Crew: David Scott, Alfred Worden and James Irwin
Landing site: Rima Hadley

Apollo 16 (April 16–27, 1972) First mission to land on the high plains.
Crew: John W. Young, T. Kenneth Mattingly, Jr., and Charles M. Duke, Jr.
Landing site: Descartes Highlands

Apollo 17 (December 7–19, 1972) Last crewed lunar mission.
Crew: Eugene A. Cernan, Ronald E. Evans and Harrison H. Schmitt.
Landing site: Taurus-Littrow Valley.

"The Blue Marble": Earth as seen during the Apollo 17 mission. This is the last picture of the Earth taken by a human being from this distance.

Pluto and Charon

Neptune

Uranus

Saturn

Jupiter

Ceres

Mars

Earth

Venus

Mercury

RELATIVE DISTANCES

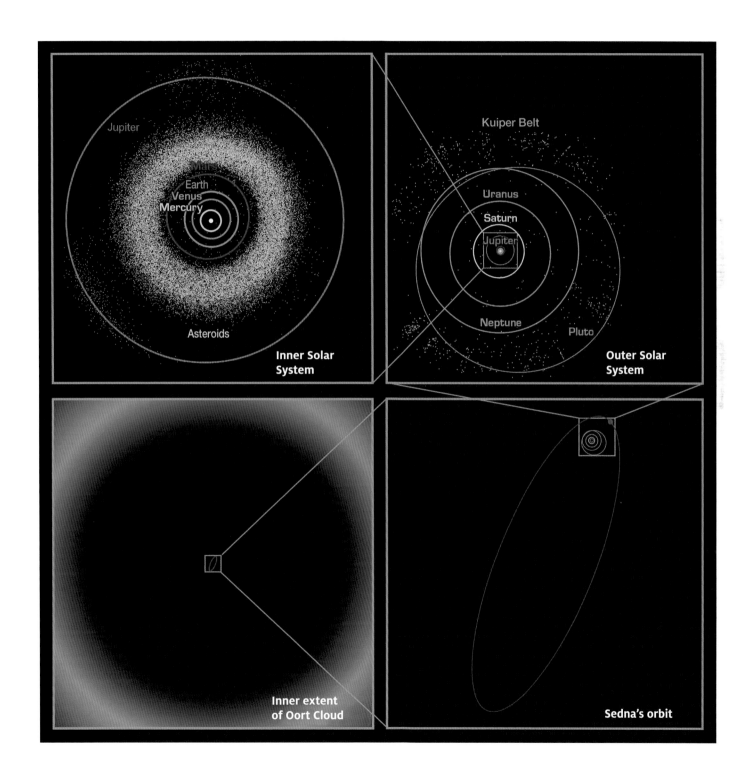

Inner Solar System

Outer Solar System

Inner extent of Oort Cloud

Sedna's orbit

APPENDIX
CHRONOLOGICAL CHART

1959-1960

DATE (launch)	MISSION	COUNTRY	DESTINATION	PAGE
January 2, 1959	Luna 1	USSR	The Moon	27
March 3, 1959	Pioneer 4	United States	The Moon	30
September 12, 1959	Luna 2	USSR	The Moon	27
October 4, 1959	Luna 3	USSR	The Moon	31
November 26, 1959	Pioneer P-3 (failure)	United States	The Moon	
March 11, 1960	Pioneer 5	United States	Sun, interplanetary space	270
April 15, 1960	Luna 1960A (failure)	USSR	The Moon	
April 18, 1960	Luna 1960B (failure)	USSR	The Moon	
October 10, 1960	Mars 1960A (failure)	USSR	Mars	
October 14, 1960	Mars 1960B (failure)	USSR	Mars	
December 15, 1960	Pioneer P-31 (failure)	United States	The Moon	

1961-1970

DATE (launch)	MISSION	COUNTRY	DESTINATION	PAGE
February 4, 1961	Sputnik 7 (failure)	USSR	Venus	
February 12, 1961	Venera 1 (failure)	USSR	Venus	96
January 26, 1962	Ranger 3 (failure)	United States	The Moon	34
April 23, 1962	Ranger 4 (failure)	United States	The Moon	34
July 22, 1962	Mariner 1 (failure)	United States	Venus	93
August 25, 1962	Sputnik 19 (failure)	USSR	Venus	
August 27, 1962	Mariner 2	United States	Venus	93
September 1, 1962	Sputnik 20 (failure)	USSR	Venus	
September 12, 1962	Sputnik 21 (failure)	USSR	Venus	
October 18, 1962	Ranger 5 (failure)	United States	The Moon	34
October 24, 1962	Mars 1962 A (failure)	USSR	Mars	
November 1, 1962	Mars 1 (failure)	USSR	Mars	
November 4, 1962	Mars 1962B (failure)	USSR	Mars	
January 4, 1963	Luna 1963A (failure)	USSR	The Moon	
February 2, 1963	Luna 1963B (failure)	USSR	The Moon	
April 2, 1963	Luna 4 (failure)	USSR	The Moon	
January 30, 1964	Ranger 6 (failure)	United States	The Moon	34
February 19, 1964	Venera 1964A (failure)	USSR	Venus	
March 21, 1964	Luna 1964A (failure)	USSR	The Moon	
March 27, 1964	Cosmos 27 (failure)	USSR	Venus	
April 2, 1964	Zond 1 (failure)	USSR	Venus	37
April 20, 1964	Luna 1964B (failure)	USSR	The Moon	
July 28, 1964	Ranger 7	United States	The Moon	34
November 5, 1964	Mariner 3 (failure)	United States	Mars	132
November 28, 1964	Mariner 4	United States	Mars	132
November 30, 1964	Zond 2 (failure)	USSR	Mars	37
February 17, 1965	Ranger 8	United States	The Moon	34
March 21, 1965	Ranger 9	United States	The Moon	34
May 9, 1965	Luna 5 (failure)	USSR	The Moon	
June 8, 1965	Luna 6 (failure)	USSR	The Moon	
July 18, 1965	Zond 3	USSR	The Moon	37
October 4, 1965	Luna 7 (failure)	USSR	The Moon	
November 12, 1965	Venera 2 (failure)	USSR	Venus	
November 16, 1965	Venera 3 (failure)	USSR	Venus	96
November 23, 1965	Cosmos 96 (failure)	USSR	Venus	

1961-1970 (continued)

DATE (launch)	MISSION	COUNTRY	DESTINATION	PAGE
November 26, 1965	Venera 1965A (failure)	USSR	Venus	
December 3, 1965	Luna 8 (failure)	USSR	The Moon	
December 16, 1965	Pioneer 6	United States	Sun, interplanetary space	271
January 31, 1966	Luna 9	USSR	The Moon	40
March 31, 1966	Luna 10	USSR	The Moon	44
May 30, 1966	Surveyor 1	United States	The Moon	46
August 10, 1966	Lunar Orbiter 1	United States	The Moon	50
August 17, 1966	Pioneer 7	United States	Sun, interplanetary space	271
August 24, 1966	Luna 11	USSR	The Moon	54
September 20, 1966	Surveyor 2	United States	The Moon	46
October 22, 1966	Luna 12	USSR	The Moon	54
November 6, 1966	Lunar Orbiter 2	United States	The Moon	50
December 21, 1966	Luna 13	United States	The Moon	56
February 5, 1967	Lunar Orbiter 3	United States	The Moon	50
April 17, 1967	Surveyor 3	United States	The Moon	46
May 4, 1967	Lunar Orbiter 4	United States	The Moon	50
June 12, 1967	Venera 4	USSR	Venus	96
June 14, 1967	Mariner 5	United States	Venus	98
June 17, 1967	Cosmos 167 (failure)	USSR	Venus	
July 14, 1967	Surveyor 4	United States	The Moon	46
July 19, 1967	Explorer 35	United States	Sun, interplanetary space	273
August 1, 1967	Lunar Orbiter 5	United States	The Moon	50
September 8, 1967	Surveyor 5	United States	The Moon	46
November 7, 1967	Surveyor 6	United States	The Moon	46
December 13, 1967	Pioneer 8	United States	Sun, interplanetary space	271
January 7, 1968	Surveyor 7	United States	Moon	46
April 7, 1968	Luna 14	USSR	Moon	54
September 14, 1968	Zond 5	USSR	Moon	37
November 8, 1968	Pioneer 9	United States	Sun, interplanetary space	271
November 10, 1968	Zond 6	USSR	The Moon	37
January 5, 1969	Venera 5	USSR	Venus	96
January 10, 1969	Venera 6	USSR	Venus	96
February 24, 1969	Mariner 6	United States	Mars	136
March 27, 1969	Mariner 7	United States	Mars	136
July 13, 1969	Luna 15	USSR	The Moon	58
August 7, 1969	Zond 7	USSR	The Moon	37
August 17, 1970	Venera 7	USSR	Venus	99
August 22, 1970	Cosmos 359 (failure)	USSR	Venus	
September 12, 1970	Luna 16	USSR	The Moon	58
October 20, 1970	Zond 8	USSR	The Moon	37
November 10, 1970	Luna 17/LK1	USSR	The Moon	62

1971-1980

DATE (launch)	MISSION	COUNTRY	DESTINATION	PAGE
May 8, 1971	Mariner 8 (failure)	United States	Mars	140
May 10, 1971	Cosmos 419 (failure)	USSR	Mars	
May 19, 1971	Mars 2 (partial failure)	USSR	Mars	144
May 28, 1971	Mars 3 (partial failure)	USSR	Mars	144
May 30, 1971	Mariner 9	United States	Mars	140
September 2, 1971	Luna 18 (failure)	USSR	The Moon	
September 28, 1971	Luna 19	USSR	The Moon	65

1971-1980 (continued)

DATE (launch)	MISSION	COUNTRY	DESTINATION	PAGE
February 14, 1972	Luna 20	USSR	The Moon	58
March 3, 1972	Pioneer 10	United States	Jupiter	212
March 27, 1972	Venera 8	USSR	Venus	99
March 31, 1972	Cosmos 482 (failure)	USSR	Venus	
January 8, 1973	Luna 21/LK2	USSR	The Moon	62
April 6, 1973	Pioneer 11	United States	Saturn	212
June 10, 1973	Explorer 49	United States	The Moon	
July 21, 1973	Mars 4 (partial failure)	USSR	Mars	146
July 25, 1973	Mars 5	USSR	Mars	146
August 5, 1973	Mars 6 (partial failure)	USSR	Mars	146
August 9, 1973	Mars 7 (failure)	USSR	Mars	146
November 3, 1973	Mariner 10	United States	Venus, Mercury	101
May 29, 1974	Luna 22	USSR	The Moon	65
October 28, 1974	Luna 23 (failure)	USSR	The Moon	60
December 10, 1974	Helios 1	United States & Germany	Sun, interplanetary space	274
June 8, 1975	Venera 9	USSR	Venus	104
June 14, 1975	Venera 10	USSR	Venus	104
August 20, 1975	Viking 1	United States	Mars	148
September 9, 1975	Viking 2	United States	Mars	148
October 16, 1975	Luna 1975A (failure)	USSR	The Moon	
January 15, 1976	Helios 2	United States & Germany	Sun, interplanetary space	274
August 9, 1976	Luna 24	USSR	The Moon	58
August 20, 1977	Voyager 2	United States	Jupiter, Saturn, Uranus, Neptune	219
September 5, 1977	Voyager 1	United States	Jupiter, Saturn	219
May 20, 1978	Pioneer Venus Orbiter	United States	Venus	108
August 8, 1978	Pioneer Venus Multiprobe	United States	Venus	111
August 12, 1978	ISEE/ICE	United States	Sun, interplanetary space	276
September 9, 1978	Venera 11	USSR	Venus	104
September 14, 1978	Venera 12	USSR	Venus	104

1981-1990

DATE (launch)	MISSION	COUNTRY	DESTINATION	PAGE
October 30, 1981	Venera 13	USSR	Venus	104
November 4, 1981	Venera 14	USSR	Venus	104
June 2, 1983	Venera 15	USSR	Venus	113
June 7, 1983	Venera 16	USSR	Venus	113
December 15, 1984	Vega 1	USSR	Venus, Halley's Comet	
December 21, 1984	Vega 2	USSR	Venus, Halley's Comet	
January 7, 1985	Sakigake	Japan	Halley's Comet	298
July 2, 1985	Giotto	Europe	Halley's Comet	301
August 18, 1985	Suisei	Japan	Halley's Comet	299
July 7, 1988	Phobos 1	USSR	Mars	160
July 12, 1988	Phobos 2	USSR	Mars	160
May 4, 1989	Magellan	United States	Venus	118
October 18, 1989	Galileo	United States	Jupiter	229, 327
January 24, 1990	Hiten-Hagoromo	Japan	The Moon	66
October 6, 1990	Ulysses	United States	Sun, interplanetary space	278

1991-2000

DATE (launch)	MISSION	COUNTRY	DESTINATION	PAGE
September 25, 1992	Mars Observer (failure)	United States	Mars	162
January 25, 1994	Clementine	United States	The Moon	67
November 1, 1994	Wind	United States	Sun, interplanetary space	282
December 2, 1995	SOHO	United States & Europe	Sun, interplanetary space	283
February 17, 1996	NEAR Shoemaker	United States	433 Eros	328
November 7, 1996	Mars Global Surveyor	United States	Mars	162
November 16, 1996	Mars 96 (failure)	USSR	Mars	161
December 4, 1996	Mars Pathfinder	United States	Mars	166
August 25, 1997	ACE	United States	Sun, interplanetary space	286
October 15, 1997	Cassini-Huygens	United States & Europe	Saturn	238
January 6, 1998	Lunar Prospector	United States	The Moon	71
July 3, 1998	Nozomi (failure)	Japan	Mars	
October 24, 1998	Deep Space 1	United States	9969 Braille & Comet Borrelly	306
December 11, 1998	Mars Climate Orbiter (failure)	United States	Mars	172
January 3, 1999	Mars Polar Lander/ Deep Space 2 (failure)	United States	Mars	172
February 7, 1999	Stardust	United States	Comet Wild 2	309

2001-2010

DATE (launch)	MISSION	COUNTRY	DESTINATION	PAGE
April 7, 2001	Mars Odyssey	United States	Mars	172
August 8, 2001	Genesis	United States	Sun, interplanetary space	288
July 3, 2002	CONTOUR (failure)	United States	Encke & Comets SW-3, d'Arrest	
May 9, 2003	Hayabusa	Japan	25143 Itokawa	332
June 2, 2003	Mars Express Beagle 2 (failure)	Europe	Mars	184
June 10, 2003	Spirit (MER-A)	United States	Mars	176
July 7, 2003	Opportunity (MER-B)	United States	Mars	176
September 27, 2003	SMART 1	Europe	The Moon	74
March 2, 2004	Rosetta	Europe	Comet 67P/C-G	313
August 3, 2004	Messenger	United States	Mercury	262
January 12, 2005	Deep Impact	United States	Comet 9P/Tempel 1	318
August 12, 2005	Mars Reconnaissance Orbiter	United States	Mars	192
November 9, 2005	Venus Express	Europe	Venus	124
January 19, 2006	New Horizons	United States	Pluto	254, 335
October 26, 2006	STEREO A & B	United States	Sun, interplanetary space	291
August 4, 2007	Phoenix	United States	Mars	200
September 14, 2007	Kaguya (SELENE)	Japan	The Moon	77
September 27, 2007	Dawn	United States	Vesta, Ceres	340
October 24, 2007	Chang'e 1	China	The Moon	80
October 22, 2008	Chandrayaan-1	India	The Moon	82
June 18, 2009	Lunar Reconnaissance Orbiter/LCROSS	United States	The Moon	86
May 20, 2010	Akatsuki/Planet C	Japan	Venus	349
October 1, 2010	Chang'e 2	China	The Moon	80

GLOSSARY

Aerobraking: a technique to modify a space-craft's trajectory by dragging it against the upper layers of a planet's atmosphere.

Altimeter: an instrument that measures both altitude and planetary surface relief.

Aphelion: the point in an elliptical orbit of a spacecraft or planet that is farthest from the center of the Sun.

Apogee: the point in an elliptical orbit of a spacecraft or celestial object that is farthest from the Earth's center.

Asteroid: a small planetoid or large meteor in orbit around the Sun.

Astronautics: the science and technology of space flight.

Astronomical Unit (AU): the average distance from the Earth to the Sun, about 150 million kilometers.

Atmosphere: a gaseous envelope around a celestial object, held in place by gravity.

Attitude: the position of an object in space relative to an external reference.

Caldera: a circular or elliptical depression of volcanic origin.

Comet: a small object in the solar system composed of ice and of dust. Comets circle the Sun, usually in highly elliptical orbits.

Cosmic rays: high-energy particles of indeterminate origin, coming from interstellar and intergalactic sources. Cosmic rays are composed of charged particles (85 to 90 percent protons, 9 to 14 percent helium nuclei) and neutral particles (gamma rays and neutrinos).

Doppler (effect): a shift in frequency of an acoustic or electromagnetic wave, observed when the distance between the source and the observer increases or decreases.

Ecliptic: the plane of the Earth's orbit that corresponds approximately to the plane of the solar system.

Exobiology: the study of the origin, structure and evolution of life forms in the solar system and throughout the Universe.

Gain: in electromagnetics, a term denoting the amplification of a signal.

Gamma rays: a type of electromagnetic radiation composed of high-energy photons (beyond 100 keV) with very short wavelengths.

Gravitational boost or assistance: the deliberate use of the gravity of a celestial body to modify or boost the direction and speed of a spacecraft's trajectory.

Gravity: the force of attraction between two bodies, proportional to their masses.

Gyroscope: a device for measuring or maintaining direction, using the principles of conservation of angular momentum.

Heliocentric orbit: orbit of an object revolving around the Sun.

Hydrazine: a liquid rocket fuel with the chemical formula of N_2H_4.

Infrared: the portion of the electromagnetic spectrum in the 780 nanometers to 1 millimeter wavelength range, intermediate between visible and microwave radiation.

Kuiper Belt: the region of the solar system extending beyond the orbit of Neptune, between 30 and 55 AU, which contains the dwarf planets Pluto, Makemake and Haumea, as well as a multitude of smaller objects.

Lagrangian Point: an orbital position where the combined gravitational pull of two large bodies results in an equilibrium where a satellite or spacecraft can rotate with them.

Launch Window: the time frame during which the position of a celestial object relative to the Earth is favorable for the launch of a spacecraft in its direction.

Light-year: a unit of distance based on the distance light travels in space in 365 days (one year). The speed of light being 299,792 kilometers per second, a light-year is equal to 9,461 trillion kilometers.

Magnetometer: an instrument that measures the strength or direction of a magnetic field.

Magnetosphere: the magnetic field of a planet or a star.

Meteorite: a celestial object that reaches Earth's surface without being completely vaporized.

Nose cone: the aerodynamic upper part of a rocket, which protects the payload or spacecraft and can be jettisoned.

Oort Cloud: a huge, spherical cloud of small objects located at about 50,000 AU from the Sun. The Oort Cloud is probably the origin of most comets, and the outer limit of the Oort Cloud forms the limit of the solar system.

Orbit: gravitationally curved trajectory of a spacecraft or celestial object around another object.

Organic Molecule: a molecule composed of at least one carbon atom linked to one or more hydrogen or other atoms.

Payload: scientific instruments and other technical equipment carried by a spacecraft for a specific mission.

Perigee: the point of an elliptical orbit of a spacecraft or celestial object that is closest to the Earth's center.

Perihelion: the point in an elliptical orbit of a spacecraft or planet that is closest to the center of the Sun.

Photopolarimeter: an optical instrument to measure the intensity and polarization of light reflected by atmospheric particles.

Pixel: the smallest element of a digital image.

Plasma: a state of matter that is in the form of charged particles (ions and electrons).

Propellant: a combination of one or more liquid or solid chemical agents used in rocket engines to propel spacecraft through oxidation-reduction reactions.

Proton: a positively charged particle that (along with neutrons) is present in the nucleus of an atom.

Radar (RAdio Detection And Ranging): an electromagnetic system used to detect an object or determine its speed by measuring the reflection of radio waves generated by a transmitter and captured by a receiver. The object's position is determined by the time interval it takes the signal to return, and the object's speed is calculated from the change in the frequency of the signal, due to the Doppler Effect.

Radioisotope: a radioactive chemical element, either natural or synthetic.

Radiometer: an instrument that measures the intensity of electromagnetic radiation at different wavelengths (ultraviolet, visible, infrared).

Regolith: a fine layer of dust on the surface of a planet, moon or asteroid, often produced by the impact of meteorites. It covers the surface of our Moon to a depth of many meters and is also present on Earth.

Revolution: the movement of a star or celestial object in a closed path around another.

Rotation: the turning of an object around its axis.

Satellite: a natural or artificial object in orbit around a more massive object.

Solar Wind: the continual flow of plasma, composed mainly of ions and electrons, which is ejected from the Sun's upper atmosphere.

Spectrometer: an instrument used to measure the properties of electromagnetic radiation (spectroscopy) or to analyze a mixture of molecules in terms of its simpler elemental components (mass spectrometry).

Sublimation: the transformation of matter from the solid state to the liquid state without passing through the gaseous state.

Sunspots: dark spots on the Sun's surface where there is intense magnetic activity, which are visible because their temperature is lower than the surrounding areas.

Telemetry: communication technology used in astronautics to measure the distance of a spacecraft and receive information from a spacecraft, mainly via radio signals.

Telluric Planet: a rocky planet with an inner structure resembling that of the Earth.

Terminator: the dividing line between the light (day) and dark (night) side of a planet or a moon.

Thermoelectric generator: an apparatus that directly transforms thermal energy into electrical energy.

Transit: the passage of a celestial object in front of a larger one: for example, Venus' transit in front of the Sun.

Ultraviolet: the portion of the electromagnetic spectrum in the 10 to 400 nanometer wavelength range, intermediate between X-rays and visible light.

Van Allen Belt: the donut-shaped zone of radiation around the Earth's magnetic equator, where high-energy particles are trapped.

AU: Astronomical Unit

CCD: Charge-Coupled Device

CNES: Centre National D'Études Spatiales

CNSA: China National Space Administration

DSN: Deep Space Network

ESA: European Space Agency

ISRO: Indian Space Research Organization

IUS: Inertial Upper Stage

JAXA: Japan Aerospace Exploration Agency

JPL: Jet Propulsion Laboratory

LIDAR: Light Detection and Ranging

NASA: National Aeronautics and Space Administration

NEA: Near-Earth Asteroid

UHF: Ultra High Frequency

VEEGA: Venus-Earth-Earth Gravity Assist

VVEJGA: Venus-Venus-Earth-Jupiter Gravity Assist

INDEX

PICTURE CREDITS

Cover: ESA/DLR/FU Berlin, NASA, D.P. Mitchell
Back Cover: Author's photograph by C. Walker
Diagram Production: Jean-Michel Girard and Bruno Lamoureux

Pages 8–9: NASA; pages 10–11: NASA, ESA; page 12: NASA; page 17: NASA, RIA Novosti; page 20: Aelfwine; page 21: RIA Novosti, NASA; pages 22–23: NASA; page 24: J-P. Metsavainio; page 25: NASA; page 26: Chris 73; page 27: Kucharek; page 28: D.P. Mitchell; page 29: RIA Novosti; page 30: NASA; page 31: NASA/NSSDC; page 34: NASA, NASA/JPL; pages 35–36: NASA; page 37: NASA, RIA Novosti; page 38: RIA Novosti, D.P. Mitchell, S. Yoshimoto; pages 39–40: RIA Novosti; page 41: ITAR-TASS; page 42: T. Stryk, D.P. Mitchell, RIA Novosti; page 43: Jodrell Bank; page 44: posting from the Russian Federation; page 45: RIA Novosti; page 46: NASA; pages 48–53: NASA; pages 54–55: RIA Novosti; page 56: NASA, D.P. Mitchell; page 57: RIA Novosti; page 58: NASA; pages 59–63: RIA Novosti; page 64: RIA Novosti, D.P. Mitchell; page 65: ITAR-TASS, D.P. Mitchell; page 66: JAXA; pages 67–69: NASA; page 70: NASA, 20th Century Fox; pages 71–73: NASA; pages 74–75: ESA; page 76: ESA/SMART-1/Space Exploration Institute; pages 77–78: JAXA; page 79: JAXA/NHK; pages 80–81: CNSA; pages 82–85: ISRO; page 86 NASA/Goddard Space Flight Center; page 87: NASA, NASA/Goddard Space Flight Center/ Arizona State University; page 88: NASA/Goddard Space Flight Center/Arizona State University; pages 89–90: NASA; page 91: Public Domain, NASA; page 92: S. Walker; pages 93–95: NASA; page 96: RIA Novosti; page 97: NASA; page 98: NASA, NASA/JPL; page 99 RIA Novosti page 100: D.P. Mitchell; page 101: NASA/KSC; page 103: NASA/JPL; page 104: NASA, ITAR-TASS; page 105: NSSDC/GSF/NASA, D.P. Mitchell; page 107: J. Whatmore/ESA; pages 108–111: NASA/JPL/USGS; page 113: RIA Novosti, NASA; page 114: D.P. Mitchell; pages 115–116: RIA Novosti; page 117: NASA/JPL; page 118: NASA/KSC, NASA/JPL; pages 119–122: NASA/JPL; page 123: public domain; pages 124–126: ESA; page 127: NASA, ESA/VIRTIS/INAF-IASF/Paris Observatory-LESIA, public domain; page 128: NASA/JPL; page 129: NASA; page 130: public domain; page 131: A. Correa; pages 132–143: NASA/JPL; pages 144–145: RIA Novosti; page 146: M. Wade; page 147: D.P. Mitchell; pages 148–156: NASA/JPL; pages 158–159: NASA/JPL; page 160: RIA Novosti; page 161: D.P. Mitchell, T. Stryk; page 162: NASA/JPL; page 163: NASA/JPL/MSSS, NASA/JPL; page 164: NASA/JPL; page 165: NASA/JPL/MSSS; pages 166–171: NASA/JPL; page 172: NASA; page 173: C. Waste/NASA; page 174: NASA/JPL/MSSS, NASA/JPL/Goddard Space Flight Center; page 175: Mamyjomarash, R. Guidice/ NASA Ames Research Center; pages 176–183: NASA/JPL-Caltech/Cornell University; page 184: J-L. Atteleyn/ESA, S. Corjava/ESA/STARSEM; pages 185–186: ESA; page 187: NASA/JPL/ASI/ESA/Rome University/MOLA Science Team/USGS; pages 188–191: ESA/DLR/FU Berlin; page 192: NASA/KSC, NASA; page 193: NASA/JPL; pages 194–199: NASA/JPL-Caltech/University of Arizona; page 200: NASA/JPL; page 201: NASA/G. Shelton, NASA/JPL; page 202: NASA/G. Fergus, NASA/JPL-Caltech/University of Arizona; page 203: NASA/JPL; pages 204–207: NASA/JPL-Caltech/ University of Arizona/ Texas A&M University; page 208: NASA/JPL/Space Science Institute; page 209: IAU/M. Kornmesser; page 210: NASA/ESA/ I. de Pater & M. Wong (UC Berkeley), NASA/JPL/ Space Science Institute; pages 211–212: NASA/JPL; page 213: E. Long/NASM; pages 215–228: NASA/JPL; page 229: NASA; page 231: NASA/JPL; pages 232–236:

NASA/JPL/University of Arizona/ University of Colorado; pages 238–239: NASA; pages 240–243: NASA/JPL/Space Science Institute; page 244: D. Monniaux, NASA, ESA; pages 245–252: ESA/NASA/JPL/ University of Arizona; page 253: Paris Observatory; pages 254–255: NASA/JHUAPL/SwRI/GSFC; pages 256–257: NASA/JHUAPL; page 258: F. Schmutzer; pages 259–260: NASA/JPL; page 261: USGS; pages 262–265: NASA/JHUAPL/CIW; page 266: SOHO/ESA/NASA; pages 270–271: NASA; page 273: NASA; page 274: ESA; pages 275–280: NASA; page 281: NASA/ESA; page 282: NASA; page 283: ESA; pages 284–285: SOHO/ESA/NASA; pages 286–287: NASA; pages 288–290: NASA/JPL-Caltech; page 291: NASA; page 292: JHUAPL; page 293: NASA/JHUAPL; page 294: T. Rector/Z. Levay/L.Frattare/STSI/NOAO/AURA/NSF; page 295: E. Weiss; page 296: ESO/S. Deiries; page 297: NASA/HST Comet Team; page 298: RIA Novosti, IKI; page 299: JAXA; page 300: NOAO/AURA/NSF; pages 301–304: ESA; page 305: Frieda; pages 306–311: NASA/JPL; page 312: NASA/

JPL-Caltech/University of Maryland/Cornell University; page 313–315: ESA; page 316: ESA/MPS; page 317: ESA; page 318: Ball Aerospace & Technologies Corp., NASA; page 319: NASA/JPL; page 321: NASA/JPL-Caltech/UMD; page 322: ESO; page 323: Lowell Observatory Archives; page 324: Lowell Observatory Archives, NASA; page 325: NASA; page 326: D. Alighieri, USGS; ThundaFunda; page 327: NASA, JPL; page 328: NASA; pages 329–330: NASA/JPL/JHUAPL; page 331: NASA; page 332: JAXA; page 334: JAXA; page 335: JHUAPL; page 336: NASA/KSC; pages 338–339: JHUAPL/SwRI; page 340: NASA/JPL; page 341: NASA/JPL, G. Shelton/NASA; page 342: McREL/NASA, NASA/JPL/MPS/DLR/IDA & Dawn Flight Team; page 343: NASA/JPL/JHUAPL/HST; page 344: JPL-Catech; page 345: ITAR-TASS; page 346: NASA/JPL page 347: P. Séguéla; page 348: NASA, ESA; page 349: JAXA; page 350: ESA, NASA/JHUAPL; page 352: ESA; page 353: NASA/JPL; page 354: ESA; pages 357–363: NASA; page 364: M.Kornmesser/IAU; page 365: R. Hurt/NASA/JPL-Caltech.

Acknowledgments

I wish to extend all my gratitude to Benoit Patar, without whom this work would never have seen the light of day and who honored me with his friendship. I would also like to thank my publisher, Michel Maillé, who lent his enthusiastic support to the project. I must not forget Gianni Caccia, Fides' artistic director, for the layout, as well as Bruno Lamoureux for his beautiful infographics and his efficiency.

I also wish to warmly thank Marc Garneau whose participation was critical to the pursuit of the work, Olivier-Louis Robert for his numerous constructive suggestions and the technical discussions that greatly enriched the book's content, and Jim Oberg for kindly agreeing to write the foreword of the English version.

I can never express enough gratitude to my wife, Corinne, for her support and her patience as my first reader, nor to my children, Maxime and Jade, for their understanding during the long months of writing, when I was even more on another planet than usual.

I need to underline the decisive role played by my friends Christian Chevrier, Régis Loisel and Marie-Hélène Loisel, who believed in my dream from the beginning and encouraged me to have my book published. In passing, a big thank you to Régis for providing the creative environment of Atelier 1606, and to Christian Hébert for discovering typos and incongruities in the text.

I cannot forget the support from the entire Montreal Neurological Institute, and in particular my close collaborators — Ariel Ase, Dominique Blais, Louis-Philippe Bernier, Gary Mo and Zizhen Zhang — and my colleague Daniel Guitton.

Finally, I wish to thank the governmental space agencies (NASA, ESA, CSNA and JAXA) for their generous policy of free access to their archives.

I dedicate this book to all the people I love – they will surely recognize themselves.

SELECTED BIBLIOGRAPHY

Books

Ball, Andrew, James Garry, Ralph Lorenz and Viktor Kerzhanovich. *Planetary Landers and Entry Probes.* Cambridge: Cambridge University Press, 2007.

Benson, Michael. *Beyond: Visions of the Interplanetary Probes.* New York: Harry N. Abrams, 2003.

Burgess, Eric. *Venus: An Errant Twin.* New York: Columbia University Press, 1985.

Burrows, William E. *The Infinite Journey: Eyewitness Accounts of Nasa and the Age of Space.* Ludlow, Shropshire: Discovery Books, 2000.

Daniels, Patricia. *The New Solar System: Ice Worlds, Moons and Planets Redefined.* Washington: National Geographic, 2009.

Fraknoi, Andrew, David Morrison and Sydney C. Wolff. *Voyages to the Planets.* Pacific Grove, California: Brooks Cole, 2003.

Furniss, Tim. *A History of Space Exploration and its Future....* Newport Pagnell, Buckinghamshire: Mercury Books, 2006.

Garlick, Mark. *Astronomy: A Visual Guide.* Richmond Hill, Ontario: Firefly Books, 2004.

Gatland, Kenneth. *The Illustrated Encyclopedia of Space Technology.* New York: Harmony, 1984.

Glover, Linda, Patricia Daniels and Andrea Gianopoulos. *Encyclopedia of Space.* Washington: National Geographic, 2004.

de Goursac, Olivier. *Visions de Mars.* Paris: Éditions de la Martinière, 2004.

Kohler, Pierre. *L'Astronomie.* Geneva: Minerva, 2006.

Kraemer, Robert S. *Beyond the Moon: A Golden Age of Planetary Exploration, 1971-1978.* New York: Penguin/Putnam, 2004.

Lang, Kenneth R. *The Cambridge Guide to the Solar System.* Cambridge: Cambridge University Press, 2003.

Light, Michael. *Full Moon.* New York: Knopf, 1999.

Moore, Patrick and H. J. P. Arnold. *Space: the First 50 Years.* New York: Sterling, 2007.

Pasachoff, Jay M. and Alex Filippenko. *The Cosmos: Astronomy in the New Millennium.* Pacific Grove, California: Thomson-Brooks Cole, 2006.

van Pelt, Michel. *Space Invaders.* New York: Copernicus Books, 2007.

Raeburn, Paul. *Mars: Uncovering the Secrets of the Red Planet.* Washington: National Geographic, 1998.

Rycroft, Michael. *Cambridge Encyclopedia of Space.* Cambridge: Cambridge University Press, 1990.

Sagan, Carl. *Pale Blue Dot.* New York: Ballantine Books, 2007.

Smith, Arthur. *Planetary Exploration : Thirty Years of Unmanned Space Probes.* Cambridge: Patrick Stephens, 1988.

Sparrow, Giles. *Spaceflight.* London: Dorling Kindersley Publishing, 2007.

Sparrow, Giles. *The Planets.* London: Quercus Publishing, 2006.

Ulivi, Paolo. *Lunar Exploration: Human Pioneers and Robotic Surveyors.* New York: Springer Praxis Books, 2004.

Yenne, Bill. *The Atlas of the Solar System.* Greenwich: Brompton, 1987.

Websites

Applied Physics Laboratory (Johns Hopkins University): http://www.jhuapl.edu/

Astro Anarchy: Astrophotographer J.P. Metsavainio's Blog: http://www.astroanarchy.blogspot.com/

CSNA (China National Space Administration): http://www.cnsa.gov.cn/n615709/cindex.html

Deep Space Probes: http://www.worldspaceflight.com/probes/outer.htm

Don Davis Space Artist and Animator: http://donaldedavis.com/

Don P. Mitchell Website: Venus, Soviet Space History, Computer Graphics, Science, Etc.: http://www.mentallandscape.com/

ESA (European Space Agency): http://www.esa.int/esaCP/index.html

JAXA (Japan Aerospace Exploration Agency): http://www.jaxa.jp/index_e.html

JPL (Jet Propulsion Laboratory of Caltech/NASA): http://www.jpl.nasa.gov/

ISRO (India Space Research Organisation): http://www.isro.org/

NASA Main Portal: http://www.nasa.gov/home/index.html

NASA Images: http://www.nasaimages.org/

Photojournal: NASA'S Image Access Homepage (JPL-NASA): http://photojournal.jpl.nasa.gov/

Red Orbit: Science, Space, Technology, Health News and Information: http://www.redorbit.com/space/

Roscosmos (Russian Federal Space Agency): http://www.federalspace.ru/

Space Daily: http://www.spacedaily.com/